STUDENT UNIT G

A2 Geography
UNIT 4

Edexcel

Specification B

Unit 4: Global Challenge (The Natural Environment)

Sue Warn

Philip Allan Updates
Market Place
Deddington
Oxfordshire
OX15 0SE

tel: 01869 338652
fax: 01869 337590
e-mail: sales@philipallan.co.uk
www.philipallan.co.uk

ISBN 0 86003 694 4

This Guide has been written specifically to support students preparing for the Edexcel Specification B A2 Geography Unit 4 examination. The content has been neither approved nor endorsed by Edexcel and remains the sole responsibility of the author.

Printed by Information Press, Eynsham, Oxford

P00129

Contents

Introduction

■ ■ ■

Content Guidance

■ ■ ■

Questions and Answers

Introduction

About this guide

The purpose of this guide is to help you understand what is required to do well in **Unit 4: Global Challenge (The Natural Environment)**. The guide is divided into three sections.

This **Introduction** explains the structure of the guide and the importance of finding linkages between units. It also provides some general advice on how to approach the unit test.

The **Content Guidance** section sets out the *bare essentials* of the specification for this half of the unit. A series of diagrams is used to help your understanding; many are simple to draw and could be used in the exam.

The **Question and Answer** section includes three sample exam questions in the style of the unit test and a cross-unit question. Sample answers of differing standards are provided, as well as examiner's comments on how to tackle each question and on where marks are gained or lost in the sample answers.

Linkages

When the specification was developed it was envisaged that students would make linkages between Units 4 and 5. There are three possible ways in which your A2 course has been organised:

- You are studying Global Challenge as your first A2 unit, for examination in January.
- You have studied Global Futures (Unit 5) before starting this unit, in which case the links to Unit 5 shown opposite are most important.
- You are not allowed to enter any examinations in January and are studying Units 4 and 5 in parallel, in which case select your Unit 5 options to provide the greatest help to your Unit 4 studies.

In the third case, you will be able to make use of the linkages shown opposite.

As you study this unit, you should make linkages wherever possible between The Natural Environment and Population and the Economy. Section C of the unit test contains questions that overarch the whole of Global Challenge.

There are links to your AS course. In Unit 1, for example, you studied small-scale ecosystems in river and coastal environments; this links to global biomes and biodiversity. Similarly, your study of the growth of millionaire cities for Unit 2 links with both international migration and globalisation of the economy.

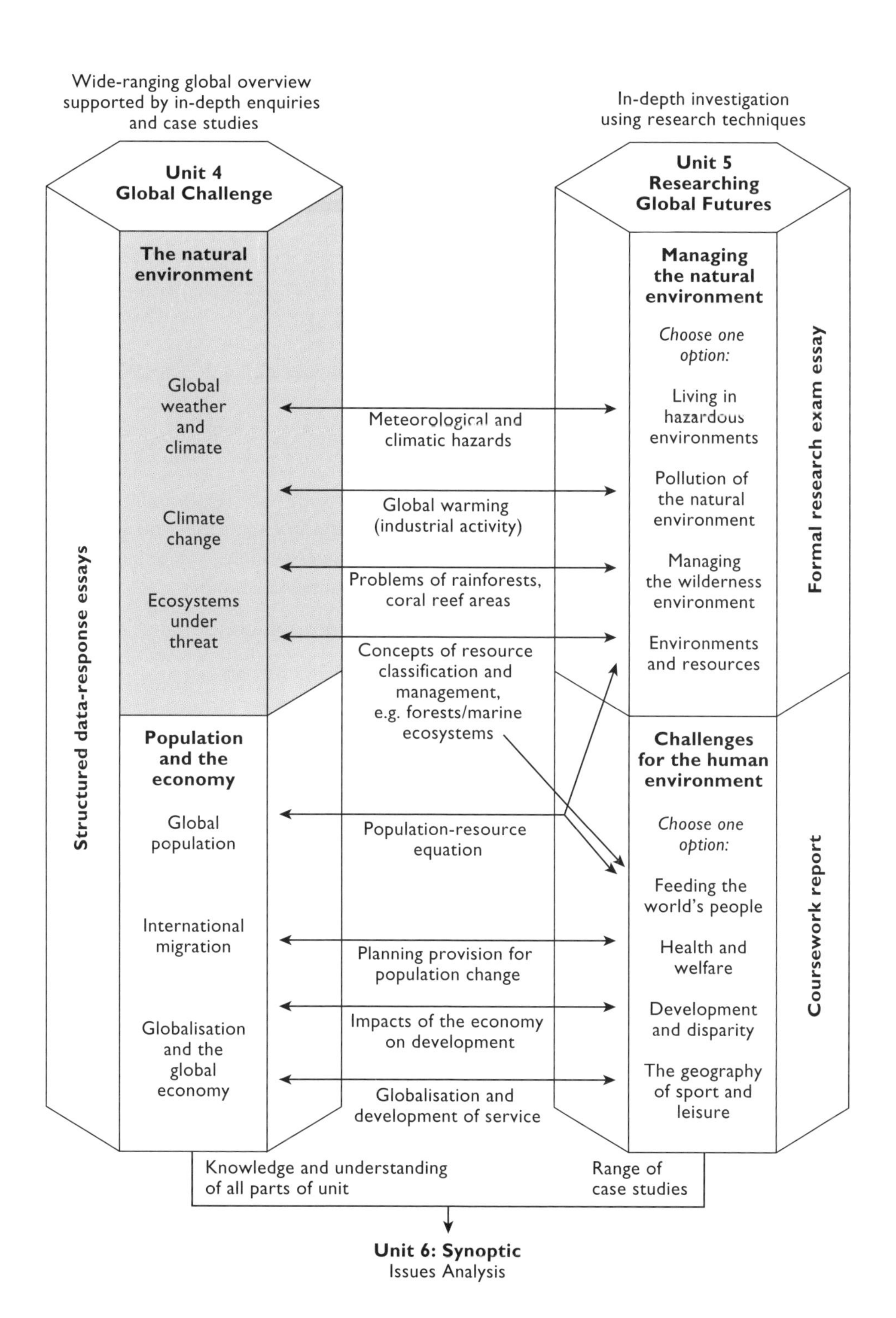
Wide-ranging global overview supported by in-depth enquiries and case studies
In-depth investigation using research techniques
Unit 4
Global Challenge
The natural environment
Global weather and climate
Climate change
Ecosystems under threat
Population and the economy
Global population
International migration
Globalisation and the global economy
Structured data-response essays
Unit 5
Researching Global Futures
Managing the natural environment
Choose one option:
Living in hazardous environments
Pollution of the natural environment
Managing the wilderness environment
Environments and resources
Challenges for the human environment
Choose one option:
Feeding the world's people
Health and welfare
Development and disparity
The geography of sport and leisure
Formal research exam essay
Coursework report
Meteorological and climatic hazards
Global warming (industrial activity)
Problems of rainforests, coral reef areas
Concepts of resource classification and management, e.g. forests/marine ecosystems
Population-resource equation
Planning provision for population change
Impacts of the economy on development
Globalisation and development of service
Knowledge and understanding of all parts of unit
Range of case studies
Unit 6: Synoptic
Issues Analysis

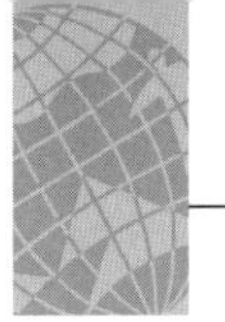

The knowledge and understanding gained in this unit will also be partially assessed in Unit 6, a synoptic issues analysis paper. Unit 6 will test your ability to draw together what you have learnt and understood, drawing on the skills that you have acquired in the context of a particular issue in an unfamiliar location or context. Issues analysis draws from all the units.

The unit test

Timing

The examination lasts for 2 hours and counts for 15% of your A2 mark. The questions in Sections A and B are each worth 25 marks whereas Section C is worth 30 marks. Therefore, you are advised to spend more time on your answer to Section C — approximately 45 minutes. You should spend 35 minutes each on Sections A and B.

Divide up your time between the three sections. Either stick to a rigid 35 minutes for A and B or, alternatively, do Section C first if you see a question that you can do. People frequently overrun on their first question, so you might find it easier to complete the higher-marked Section C question in the first 45 minutes.

Section B is covered in the companion guide, *Unit 4: Global Challenge (Population and the Economy)*.

Concepts, theories and geographical terms

At A2, the skill of using the correct vocabulary is essential. It is a good idea to compile a list of key terms as you meet them. Use your textbooks and your class notes to build up your own dictionary.

Key ideas and theories are more important at A2 than at AS, and they will be identified for you in this book using **bold type**. These are often conflicting, and you should appreciate these differences.

This unit enables you to make linkages between weather/climate and climate/ecosystems. When you are researching world issues such as climate change, support your work with examples from areas at all states of development along the continuum LDCs → LEDCs → NICs → MEDCs.

This unit is about broad issues that the world faces. The diagram opposite summarises some of the key ideas. Because this is an A2 unit, you are expected to read around these major issues and be aware of the concepts and theories that underpin them.

Global Challenge is a very big unit. Undoubtedly you will become more interested in some issues than others, but because of the way that the examination paper is organised, you are strongly advised *not* to exclude any element from your studies. For instance, if you leave out climate or economy, you could struggle on the cross-unit questions in Section C.

The topics are often current issues in the world news, such as asylum seekers, global warming, threats to forests or grasslands or coral reefs, the debt crisis and the global economy. You can help yourself to succeed by reading quality newspapers and articles in magazines such as *Geography Review*, *New Scientist* and *The Economist*. However, you still need to read textbooks and should not rely on one book.

Examination technique

The main skill that you need is that of writing semi-structured and open-ended essays. Questions are normally in two parts, with the occasional three-part question. The examination paper is accompanied by a resources booklet. This contains a wide variety of resources for you to interpret and use as a stimulus for your answers. These include:

- satellite, vertical or oblique aerial and ground-level photographs
- maps at various scales
- articles and cartoons from newspapers
- tables of data
- synoptic charts — weather maps

Data require close analysis because they often contain clues to your answer. The skills that you require are summarised below:

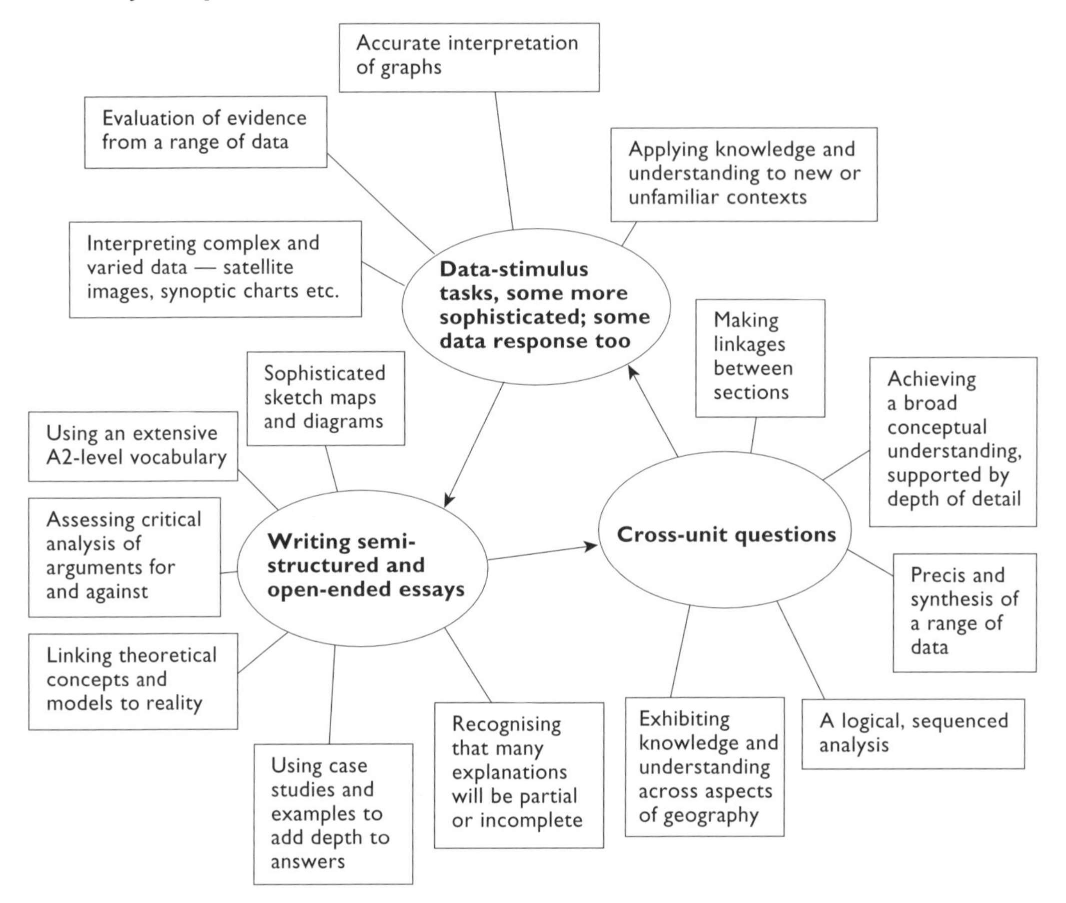

Choice of questions

You must answer *three* questions, one from each section (each containing three questions).

- **Section A: The Natural Environment**. The examiners will ensure that at least one question covers weather and climate and another is on global biomes and biodiversity.
- **Section B: Population and the Economy**. There will always be a question on population and migration and one on the global economy.
- **Section C: Cross-unit** questions.

The examiners will try to ensure coverage right across the specification. If a topic is not included in Section A or B, it is probable that it will be a part of Section C. Do remember that the same topic cannot be set every year and it is most probable that the topics set in the previous examination will not reappear.

There are three stages to selecting questions:

Stage 1
Make sure that you understand the resource and do not be seduced by it!

Stage 2
Look carefully at the knowledge and understanding required and check that you can illustrate your work with appropriate examples and, where required, an in-depth case study. In your first 5 minutes, read through the paper and list your learnt case studies on the question paper so that you know which questions are the best choices. Selecting questions from Sections A and B is normally easier, so it is probably a good idea to attempt these sections first, but *following the strict timing framework suggested on page 6*. The framework does allow more time for planning and developing the answer to Section C, which carries more marks.

Stage 3
Draw up a simple plan for each sub-section that reflects the subject matter of the question, paying attention to the command words.

Managing questions

Command words

Your question management will need to be more sophisticated than at AS. The resources are there to stimulate your responses. Describing a map, diagram or graph will not be enough because the questions will probably ask you to explain or give reasons for what is shown.

Some command words are more demanding at A2. You will often be asked 'To what extent', 'Critically assess', 'Evaluate', and 'Assess the factors that'. These are evaluative questions which anticipate that there is more than one explanation. You are expected to say which factors or explanations are more important and why. You will also be asked to 'Discuss the assertion', a command that expects a discussion of both sides of a case. More questions will expect you to 'Explain', for which description is not enough. 'Analyse' often implies some numerical analysis.

Use of diagrams and maps

It is always worth learning relevant and easy-to-draw diagrams. In this guide, there are many diagrams which you will find very useful, for example when looking at the spectrum of development. Maps and diagrams can save words provided that you integrate them into your answer and summarise the key points. Summary tables may also be used and can be effective.

Time management

Do remember to manage your time (see page 6). Missing a section of a question will probably lose you one grade on the paper.

Only spend a limited time planning. Allow a maximum of 10 minutes for question selection and planning time (5 minutes per question) for Sections A and B. Section C has more time allocated but here it is expected that you do need around 8–10 minutes of thinking time. Although the questions are each worth 30 marks, this reflects the challenge of Section C questions; you will not necessarily have to write more.

Approximate guide to the time and length of answers

Mark	Time in minutes (Sections A and B)	Length in words (approx.)	Time in minutes (Section C)	Length in words (approx.)
18/20	25	700	30	800
15	18	600	20	650
12/13	17	500	15	500
10	12	350	10	350
5	5	200	5	200
Planning time	Up to 5 minutes per question		8–10 minutes	

Note: relevant diagrams and maps may take up to 20% off the time and length figures.

Quality of written communication

Four marks on this paper are added at the end for the quality of written communication. These are extra marks and are awarded for:

- the structure and ordering of your response into a logical answer to the question, using appropriate paragraphing (single-paragraph answers will not score the maximum)
- the appropriate use of geographical terminology
- your punctuation and grammar
- the quality of your spelling (if you are dyslexic, seek special consideration)
- introducing and concluding the sections; these should not be too long, because a quality main body to an answer is required

Handling cross-unit questions

Cross-unit questions allow you to show how you can use information from across Unit 4. Hopefully, as you have studied this unit, you have tried to make linkages between The Natural Environment and Population and the Economy.

Do not forget, you can use knowledge and understanding and examples from anywhere in your course, so you may be able to use some of your research from Unit 5, *especially* if you studied pollution, wilderness and development.

These questions are very important because:

- they carry 30 marks (30/84)
- they tend to require more planning and thought before you start writing. Allow up to 10 minutes to plan out your answer, leaving 35 minutes' writing time.
- they might appear more challenging, so you need to make a wise choice and decide carefully which case studies you are going to use to support your answer.

Content Guidance

The **Natural Environment** sub-unit of Unit 4 comprises five sections:

- **introduction**
- **atmospheric processes**
- **introducing biomes**
- **global ecological causes and management of threats**
- **global ecofutures**

These five sections form the basis of Section A of Unit 4, but the concepts, theories and ideas that are the backbone of **The Natural Environment** are equally applicable to the cross-unit questions (Section C).

Throughout the Content Guidance, key terms are in **bold**. This should help you to build up your own dictionary of essential terms.

Introducing people, weather and climate

The difference between weather and climate can be expressed simply as 'weather is what you get; climate is what you expect'.

- Weather is the day-to-day changes in atmospheric conditions, at any given time or place.
- Climate is the average of weather conditions at a specific place over 30 or more years.

In most areas of the world, distinct climate trends emerge. However, the UK and most other mid-latitude areas have some of the most changeable climates, so it is hardly surprising that the distinction between weather and climate is often blurred.

Impact of weather and climate

Weather and climate influence the environment, the economy and people, especially when the pattern deviates from the expected. Economic activities are particularly sensitive to extremes, in which the weather becomes hazardous.

Impact of weather

- **Insurance** — severe weather increases accident risk, leading to widespread claims.
- **Energy** — severe winter weather increases energy use and cost. Drought conditions can lead to problems for hydroelectric power.
- **Building** — dry weather allows maximum building work, but subsidence increases. Severe weather increases damage. Cold weather can lead to frost damage.
- **Agriculture** — too much or too little rain causes problems.
- **Retailing** — wet weather sells umbrellas and anoraks. Dry, hot weather shifts ice cream, beer and bikinis.
- **Sport, leisure and tourism** — hot, dry, sunny conditions promote tourism, as does the incidence of snow for skiing. Cold, marginal conditions increase the risk for outdoor leisure. Unseasonable heat waves lead to a British resort boom.
- **Transport and communications** — roads are disrupted by snow. High winds bring down overhead railway cables.
- **Health risks** — asthmatics experience problems in fog, smog and anticyclones.

Impact of climate

- **Biodiversity** is influenced by climate as plant productivity is controlled by heat and temperature.
- Animals and plants **adapt** to particular climates.
- **Water supplies** are strongly influenced by climate, as is water demand.
- Impacts on **landscape** processes include frost shattering.

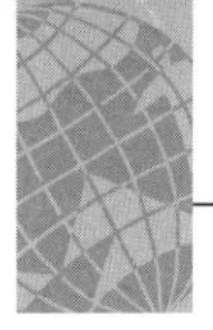

- **Disease** spreads rapidly in hot, wet conditions.
- **Wellbeing** is affected. Some people suffer from seasonal affective disorder (SAD), especially in polar areas.
- **Seasonal changes** influence migration patterns, for example of nomads.

Climate is long-term and more predictable than weather. However, seasonality has profound effects on human wellbeing and economic activities.

Environmental determinism suggests that people and their activities are determined by nature. Some environments are more appropriate to live in than others, because their climates provide food security and freedom from killer diseases. However, there are numerous examples, often in LEDCs, where people work *with* the variations in climate for maximum economic advantage.

Other scientists argue in favour of the **technological fix**. Technology:

- creates new solutions to overcome environmental constraints affecting economic growth
- manages environmental problems, such as drought or excessive frost

As with weather, deviation from the established norm causes problems. Humans can only afford the cost of working round nature's limitations when these are predictable and only cope with the unexpected when necessary.

Weather forecasting

Data collection and analysis

Over 10 000 **surface stations** collect data at 0000, 0600, 1200 and 1800 GMT. These include:

- weather ships and automatic buoys at sea
- manned and automated stations on land

In the **air**, radiosondes and aircraft collect data in the upper air while vertical wind profilers and radar (including improved radar for tracking small-scale systems such as tornadoes) monitor changes lower down.

In **space**, polar satellites (NOAA), geostationary satellites (meteosat), and geostationary and operational environment satellites (GOES) are able to measure and track systems.

The data are transmitted to weather centres, of which there is a global hierarchy. Information is collated, signals and satellite data decoded and the current data input into computer models. A **synopsis** is produced.

Forecasting the weather

There are different types of forecasting:

- Synoptic weather forecasting, using surface and upper air data to predict system movement.

- Numerical and statistical forecasting, using analogue methods that compare current trends with past records.
- Ensemble forecasting, using a variety of computer models with the input of slightly different starting conditions to predict the likelihood of, for example, snow.

The forecast is communicated to a variety of users. There are public forecasts, for example by the BBC, commercial weather forecasts used by industry and commerce, and a subscription service for aviation.

The type of forecast needed depends on the user's requirements. Forecasts can be:
- current — tornado watch, severe thunderstorms
- short-range — beach trip, road gritting
- medium-range — farmers for lambing management
- longer-range — retailers launching winter collections, organisers planning outdoor events, insurers

These are listed in order of decreasing accuracy.

Refining the forecasting process

This involves evaluation of the forecasting, including the accuracy and clarity of the data. It is based on feedback provided by forecasters and a cost–benefit analysis of the advice carried out by users.

Trends in weather forecasting

The increasing interest in weather data interpretation by government and commercial organisations may be linked to concerns over increased unpredictability, greater occurrence of extreme weather events and the possible impacts of short-term climate change.

The increased reliability of weather forecasting is due to improved understanding of atmospheric processes and advances in technology for data collection and analysis. For example:
- new-generation satellites provide higher resolution pictures and can begin recording temperature or humidity. Remote sensing is essential for better coverage.
- new-generation radar includes upward-pointing Doppler units which provide almost continuous wind profiles from the ground to the upper troposphere. This gives greater detail on **jet streams**, which are an important influence on surface weather. The latest radar and new computer models can also track small-scale systems, such as tornadoes, providing accurate 'now' forecasts.
- new-generation computers allow speedier numerical analysis of analogue weather data and more sophisticated modelling, at economic cost.

The result is a high success rate for short-term forecasts, and the increasing utility of longer-term forecasts. Cost–benefit analyses are also increasingly favourable. Higher quantities of quality data are collected and lower operating costs (automation, speed of analysis) produce more benefits for users of meteorological services, who in turn are prepared to pay more.

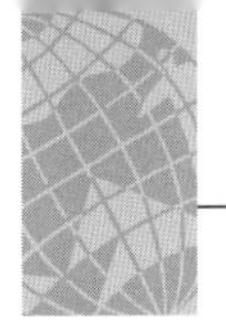

Effective forecasting involves:
- predicting the type, timing, and magnitude of an event correctly
- influencing decision-makers to make an appropriate response

Forecast failures

Weather forecasting is inexact and forecasters can get it wrong (see Table 1).

Table 1

Reason for failure	Examples	Possible solutions
Lack of data — both absolute shortage and infrequent synoptic observations	Early hurricanes, such as Galveston in 1900, with nearly 8000 deaths	• Improve data monitoring network — National Hurricane Tracking system established in Miami • Develop better hurricane monitoring technology, such as new satellite imagery
Lack of understanding of data and their significance	This was a partial factor in the October 1987 hurricane in which a mid-latitude depression intensified unexpectedly over the Bay of Biscay and tracked across southeast England causing winds of 100 mph (because of a severe pressure gradient)	• Improve data modelling • Err on the side of caution — a system of severe weather advisory warnings now operates • Improved upper air readings
System develops in a different way from the prediction in terms of timing, tracking or intensity	October 1987 hurricane was expected to track up the English Channel and be less intense	• Improve computer modelling to provide better warnings • Use radar to map small-scale systems • Better upper air data, using wind profiles to provide fuller details of jet stream patterns
Marginal conditions change very slightly	Problems with forecasting either fog or snow; it only takes a slight change of wind, humidity or temperature (about 0.5°C)	• Use less precise language to infer possible un-predictability • Talk about probability in % terms. • Ensemble forecasting to model a range of possibilities
Forecast correct, but communication or action fails	Some LEDCs did not believe the predicted severity of the 1998 El Niño event and failed to prepare fully	• Improve warning procedures • Educate decision-makers and provide aid

Atmospheric processes

Changing weather in mid-latitude areas

It can be argued that in places like the UK, Scandinavia, and northwest Canada, people only experience *weather* rather than having a climate. The changing weather in these places is a result of a number of factors, including:

- location at a meeting point of polar and tropical air streams, so they experience very contrasting airflows
- considerable periods of rapidly moving cyclonic weather, brought about by sequences of depressions
- marked seasonality, resulting from the migration of the heat equator
- stable anticyclonic conditions occuring only around 10% of the year

Rossby waves and jet streams

A key to understanding the changing weather is to think *three-dimensionally*. Surface activity has an impact on upper air activity and is also strongly controlled by it.

A belt of upper-air westerly winds occurs vertically above the polar front. It forms because of marked temperature contrasts, which result from the convergence of tropical and polar air. Temperature contrasts at the surface produce marked pressure gradients in the upper air, which in turn lead to extremely strong winds. These upper-air winds follow meandering paths called **Rossby waves**. The pattern of Rossby waves varies seasonally, with four to six waves in the summer and three in winter, in a continuous belt around the globe in mid-latitudes. Within the upper westerly wind belts there are narrow bands of extremely fast moving air (usually around 150–200 km per hour), known as **jet streams**.

The **polar front jet stream** is the key to understanding changing surface weather because it controls the formation, survival and decay of lower-level weather systems. This is illustrated in Figure 1 (page 18).

The **coriolis** effect spins the jet stream round the globe. Air travels through the Rossby waves at varying speeds, going slower as it turns away from the poles. At X on the diagram, excess air piles up because it is constricted. The convergence, east of the upper air ridge, leads to some air being forced down. This causes a high-pressure area (anticyclone) at the surface. As the jet speeds up from the upper air trough, it bends back towards the poles, diverges and spreads at Y. Here, air is sucked up from the surface, creating a low (depression).

Seasonally, the position of the polar front changes. It is usually further north in summer. Occasionally in winter, it occupies an anomalous position very far south over central Europe. Depressions tend to track along the polar front, so there are fewer in summer and more in autumn and winter. As the polar front moves, so does the position of the polar front jet stream aloft.

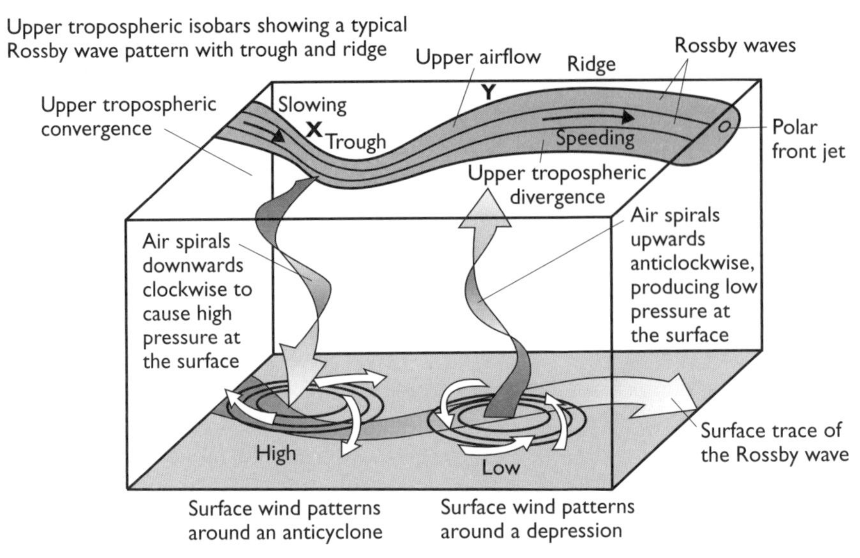

Figure 1 The relationship between upper convergence and divergence and surface anticyclones and depressions (northern hemisphere)

Over time, the Rossby waves change position (Figure 2(a)), as does the jet stream, which is a faster moving core within them. A strong jet stream results in a mobile pattern of west to east weather.

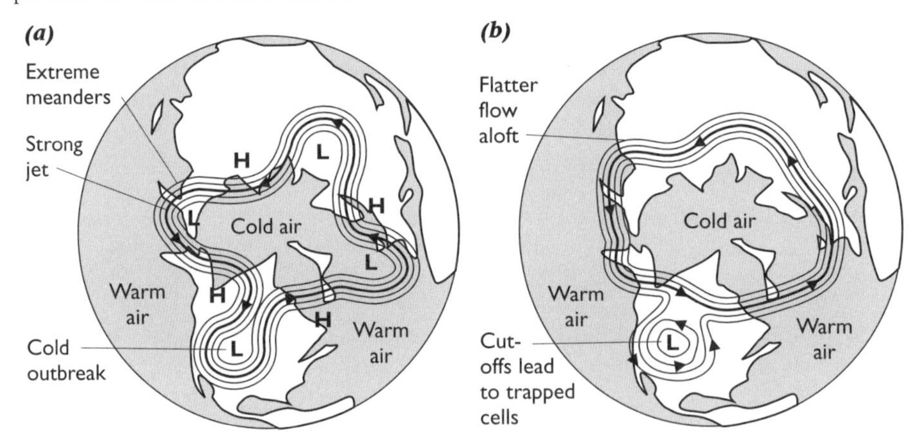

Figure 2 (a) Strong waves form in the upper airflow (b) Return to a period of flatter flow aloft

A weaker jet stream allows more incursions of cold air southwards (or warm air northwards), giving rise to more extreme weather conditions, with the probability of blocking anticyclones and deep, stationary lows. In time, the redistribution of air leads to a return of flatter flow aloft (Figure 2(b)), with less stormy weather at the surface.

Weather types

The UK is under the influence of seven general types of weather. Five of these result from airflows and two result from systems. The direction that an airmass moves in an airflow is determined by the location of pressure systems. Air always flows from high to low pressure, with clockwise circulation around a high and anticlockwise circulation around a low in the northern hemisphere.

Table 2 How Lamb's weather types affect the weather

Lamb's weather type	Occurrence	Likely airmass	Characteristic weather
Westerly, northwesterly	20%	Polar maritime, Pm (both tracks)	Cool, changeable conditions, strong winds, frequent showers on exposed western coasts
Northerly	7%	Arctic maritime, Am	Cold weather at all seasons, snow or sleet showers in the north
Northeasterly, easterly	8%	Polar continental, Pc (winter only)	Cold in winter, snow showers to east-facing coasts, but generally dry
Southeasterly	15%	Tropical continental transitional, Tct (central Europe, summer only)	In summer, a different airmass brings warm, dry weather; sea fog on coasts
Southerly	5%	Tropical continental, Tc	Very warm, dry, sometimes heat-wave conditions Some convectional uplift — thunderstorms May lead to late Indian summer
Southwesterly	20%	Tropical maritime, Tm	Very mild, damp weather in winter, especially in southwest Generally warm, thundery, unstable weather in summer
Cyclonic	15%	Formed from contrasting air-masses, usually Tm/Pm	Changeable, fast-moving, rainy conditions accompanied by strong winds Usually depressions pass in 24 hours but occasionally a deep depression stays, leading to widespread flooding
Anticyclonic	10%	Associated with continental air-masses	Very settled, stable conditions Note the contrast between winter highs (intense cold, frost and fog) and summer highs (hot and dry)

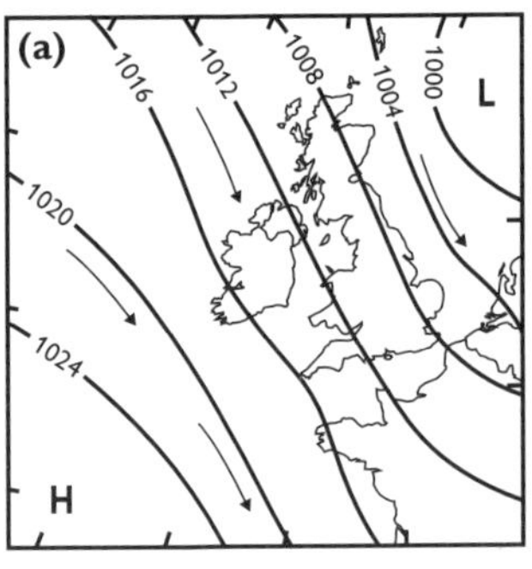

(a) The westerly and northwesterly patterns
After the passage of an eastward-moving cold front, air is drawn from the northwest Atlantic. The air is cool and picks up moisture as it crosses the ocean (Pm)

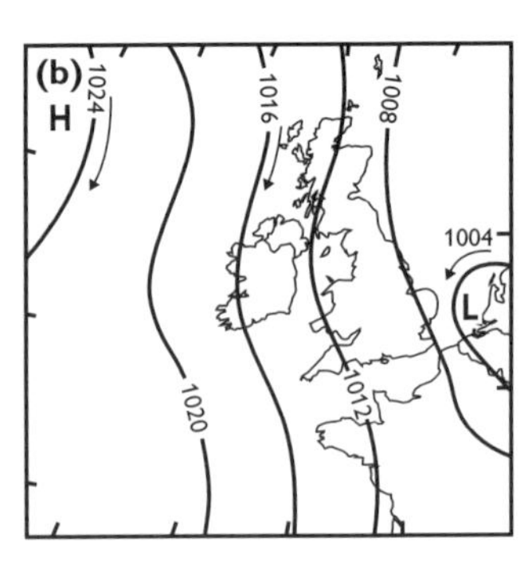

(b) The northerly pattern
When pressure is low over the UK and North Sea, air is drawn in from the Icelandic high, in a very cold northerly air stream, especially in winter (Am)

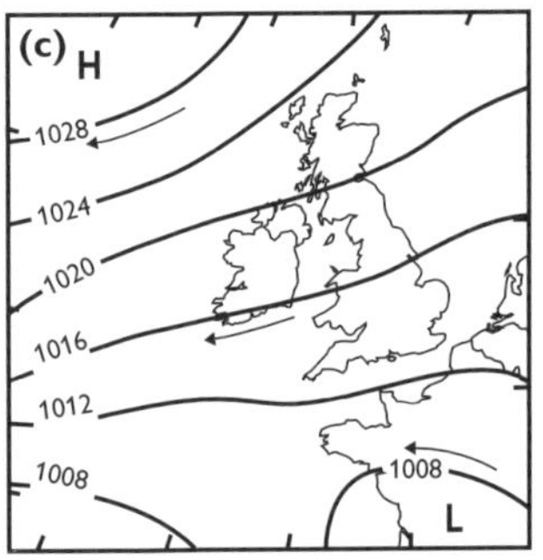

(c) The northeasterly and easterly patterns
A low pressure forms in the continents and a high over Scandinavia or the North Atlantic, resulting in a northeasterly or easterly cold airflow from Siberia, usually only in winter (Pc)

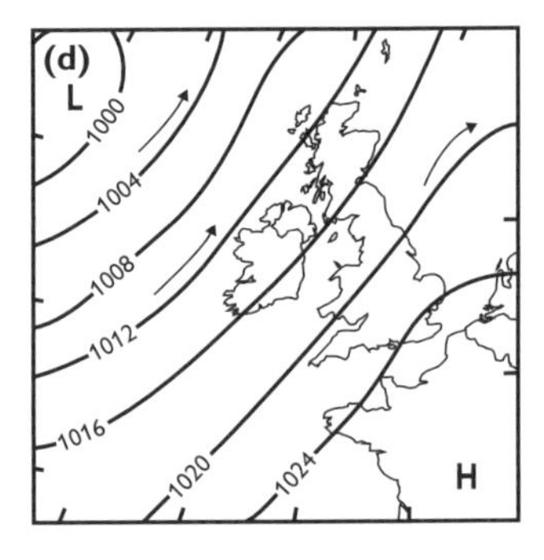

(d) The southwesterly pattern
Depressions tracks northeastward between Scotland and Iceland and draw in tropical moist air from the Azores in the southwest (Tm)

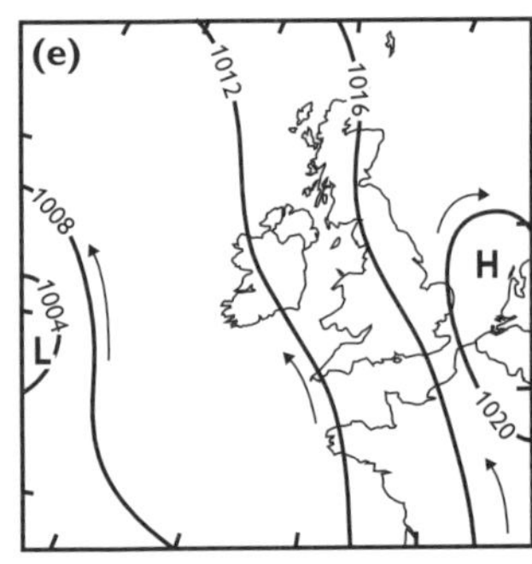

(e) The southeasterly and southerly patterns
These rely on a high pressure forming over the continent. Air is drawn from the south towards an Atlantic low. Southeastern air is warm — comes from Danube (Tct). Southerly air is very warm and dry — comes from Sahara (Tc)

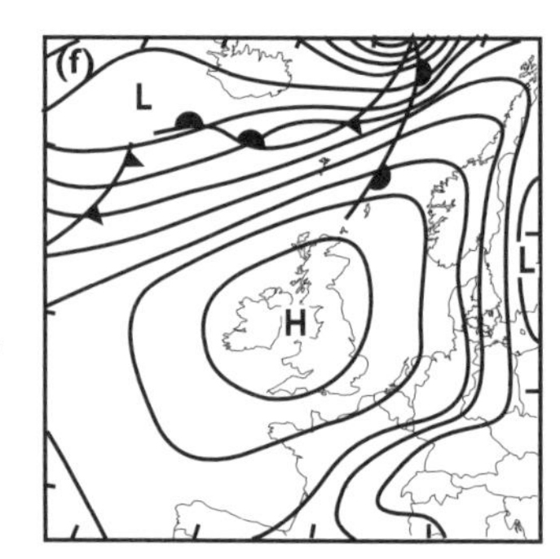

(f) No dominant airflow
Anticyclonic weather leads to very stable conditions as blocking highs develop, leading to:

- cold, clear, frost weather in winter (Pc)
- warm, sunny, almost heatwave conditions in summer (Tc)

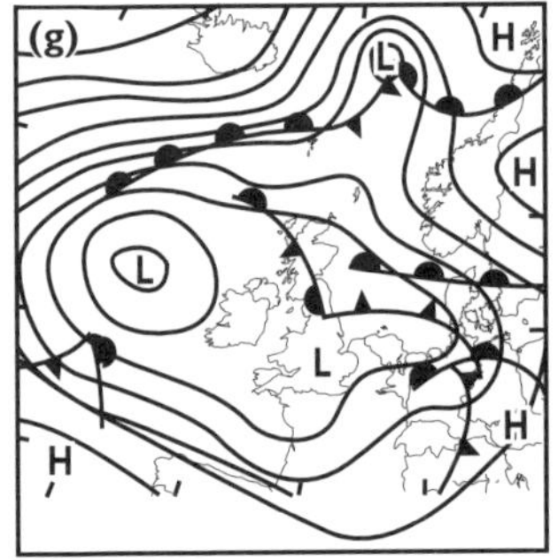

(g) No dominant airflow
Cyclonic weather occurs at the polar front where tropical airmasses meet polar airmasses (usually Tm/Pm) to create a low pressure system. Each one takes 1–2 days to pass but may take a whole week, when a series of depressions leads to very changeable weather

Figure 3 Seven weather types that affect the UK (Lamb's weather types)

Airmasses

An airmass is 'a body of air of uniform nature over a large area'. Airmasses form in stable areas such as polar and tropical high-pressure belts. The source determines the basic characteristics of airmasses:

- Tropical airmasses are warm.
- Polar airmasses are cold.
- Maritime airmasses are moist.
- Continental airmasses are dry.

As airmasses flow away from their source region, their character changes because they pass over warmer or colder surfaces. This affects the degree of stability of their lowest layers and the picking up of moisture when crossing water. Therefore, the impact of airmasses reaching the UK is a result of not only their source but also the track they have taken. For example, polar maritime airmasses can take either a direct track or a 'returning' southerly track, which brings warmer temperatures and is more stable. The airmasses have a different impact on the weather in winter and summer as some are more dominant in particular seasons. Figure 4 and the text that follows summarise their impact on UK weather.

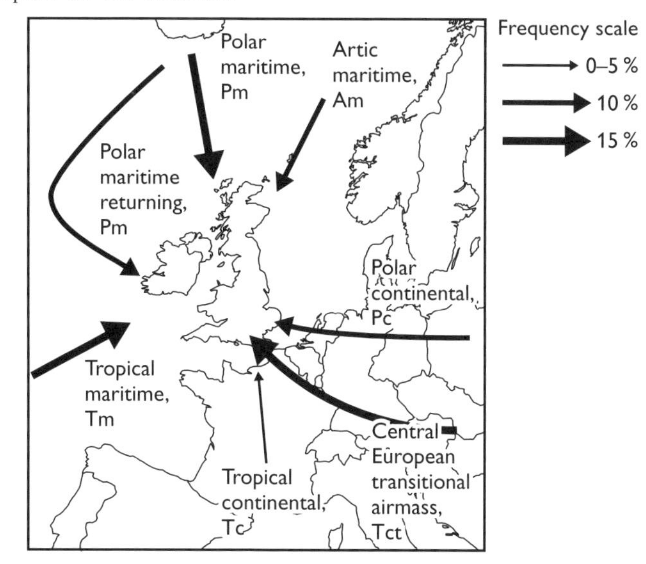

Figure 4 The impact of airmasses on UK weather

- **Polar maritime (Pm)**: the source is Canada or the north Atlantic. It is warmed by oceans in winter as it tracks southwards, so is moist and unstable. It leads to cool, showery winter weather, 6–8°C. Visibility is usually good. Cumulus clouds form. In summer, the air is damp and cool, 10–12°C. It leads to cool, showery conditions, with bright, clear conditions between.
- **Returning track polar maritime (rPm)**: it tends to be moist, but produces warm conditions (14°C in summer; 16°C in winter) and in many ways is transitional to Tm. It is a common airmass.

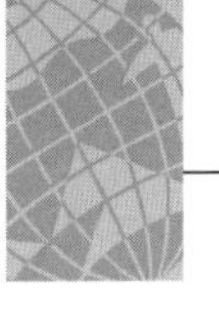

- **Arctic maritime (Am)**: the source is the icy seas off Scandinavia. It is extremely cold. As it moves south, the lowest layers are warmed by the north Atlantic; it becomes unstable and picks up moisture. In winter, conditions are extremely cold with blizzards in northern Scotland. Very rare in summer, it brings very cool, showery conditions with good visibility.
- **Polar continental (Pc)**: the source (in winter only) is the Siberian High. There, rapid cooling leads to very low temperatures (–10°C). It brings very cold, dry air to the UK (–1–0°C). As it crosses the relatively warm North Sea, it is warmed and picks up some moisture. Therefore, it can bring snow showers to hilly areas in eastern England. The wind chill factor exaggerates the coldness but little rain occurs because of relatively low amounts of moisture in the airmass.
- **Transitional (Tct)**: the source in summer is the Central Europe High, e.g. the Danube. It brings warm, dry, sunny weather. As it passes over the North Sea, it picks up moisture, but is cooled and so is stable. It leads to coastal sea fog and high pollution levels. Thunderstorms can occur as it reaches warmer land (convectional uplift).
- **Tropical maritime (Tm)**: the source is the high-pressure Azores area. The airmass is moist as it crosses the ocean but as it moves north, it is cooled and is therefore stable. It brings mild, damp, foggy conditions in winter (11–12°C), 'muggy' with stratus clouds. If forced to rise over hills, heavy rainfall occurs. It is frequently a warm airmass in a depression, so frontal rainfall also occurs. In summer, the air is warm (22–23°C) and moist. The air is cooled as it moves north, so dull, drizzly weather can result in western England. Sea fog is common in coastal regions.
- **Tropical continental (Tc)**: the source is northern Africa (the Sahara). It is uncommon in summer, and extremely rare in winter. It produces summer heat waves from the hot, dry source; air temperature up to 30°C. It is a stable airmass, cooled as it tracks north. When it reaches warmer land, it may lead to convectional uplift and thunderstorms, coastal fog on the south coast and poor visibility. In autumn, it can lead to unseasonably high temperatures.

Weather systems

Depressions

Depressions are low-pressure systems. In mid-latitudes such as the UK, they occur when pronounced Rossby waves lead to a strong jet stream in the troposphere, which generates marked activity along the polar front. Depressions begin to form when a tropical airmass meets a polar airmass, forming a vortex of anticlockwise swirling air.

- Every year, on average, mid-latitude areas such as the UK experience about 50–60 depressions.
- For 15–20% of the year, the cyclonic weather type prevails.
- On average, each depression lasts 3–5 days from formation to occlusion and decay **(life cycle)**.
- The combination of changes over time combined with rapid movement leads to a pattern of rapidly changing weather, which follows a clear sequence (Figure 5).

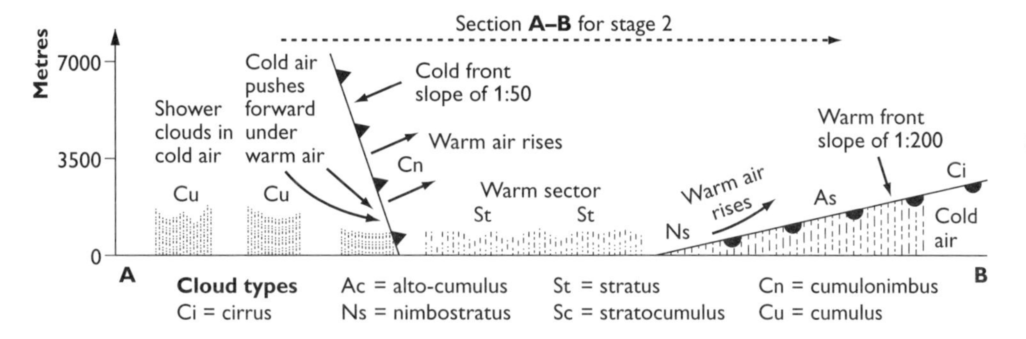

Figure 5 The passage of a depression

Development and movement of low-pressure systems

The development and movement of low-pressure systems are controlled by the polar front jet stream. The rate the air is sucked upwards determines how low the surface pressure is. A very marked low produces a deep depression with tightly packed isobars and strong winds. The speed of the jet stream within the upper-air westerlies determines the rate of movement — the **passage of the depression**.

The stage of a passing depression (see Figure 6) influences the nature of its impact. Further variation is provided by the nature and type of converging airmasses that form the system and the temperature difference between them.

Severe weather events

In general, the rapidly changing weather associated with depressions is a bonus. The equable climate with abundant well-distributed rainfall and the rapidly moving air leads to an absence of severe fogs. However, occasionally depressions can lead to very **severe weather events.**

- The most common hazardous impact is high winds and gales. These are caused by the steep pressure gradient brought about by the temperature difference between converging polar and tropical airmasses. Gales have a huge impact on infrastructure (e.g. Burns day storm in January 1990).
- Coastal floods result from onshore gale-force winds being drawn towards a deep depression, which leads to a storm surge developing, as water is sucked up. When this is combined with high spring tides, huge breakers are driven downwards, breaking sea defences and causing localised flooding (e.g. Towyn, February 1990).
- Violent thunderstorms are associated with uplift at the cold front, leading to the development of cumulonimbus clouds. Cold northwesterly or northerly air wedges underneath warm air. Occasionally violent hailstorms occur (e.g. London, 1968).
- Static depressions, usually controlled by Rossby wave patterns, can cause heavy rainfall because the warm air within them (tropical maritime airmass) can hold large quantities of moisture. This can lead to localised flooding (e.g. Lynton, 1952).
- A succession of depressions can lead to high levels of antecedent moisture in the ground, so that even modest rainfall leads to flooding (e.g. Midlands, April 1998 and November 2000).

- Snow, especially in areas of high altitude, occurs when the land surface has become very cold after a blocking winter anticyclone. When warm Atlantic air finally comes via a depression, it is forced to rise, and precipitation is initially in the form of snow. Sometimes, a deep depression draws in very cold arctic and polar air as it tracks northwards, leading to particularly heavy snow (e.g. American 'storm of the century', March 1993).

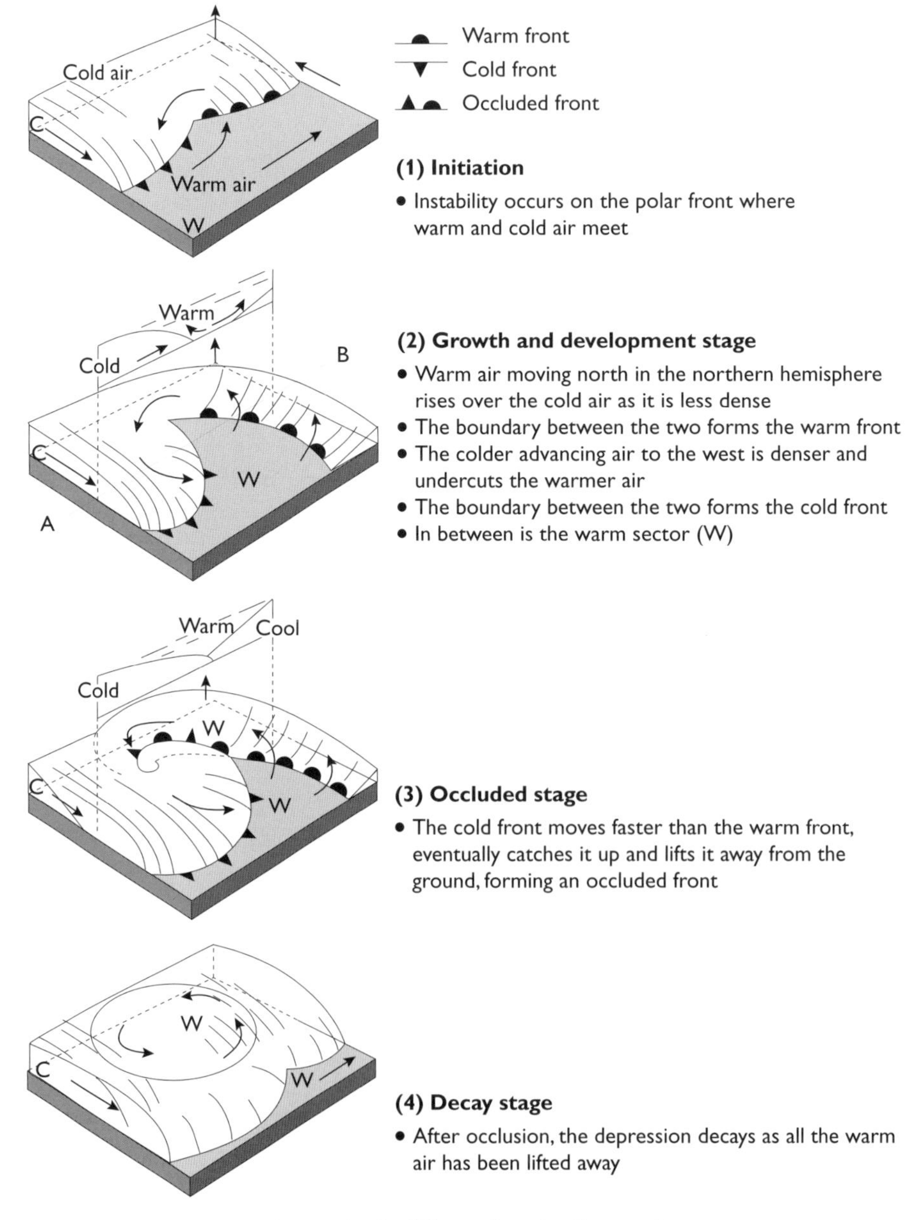

Figure 6 The life cycle of a depression

Anticyclones: a potential hazard?

An anticyclone is a large mass of subsiding air, which causes high pressure at the Earth's surface. Globally, high-pressure systems are concentrated at the poles and in the subtropical high-pressure belt between latitudes 25°–35° north and south. The main source regions of high pressure, which affects the UK for less than 10% of the year, are Siberia and Scandinavia in winter, and central and southern Europe in summer. Winter and summer anticyclones have many common features because of their stability and the subsiding air at the centre, which is adiabatically warmed, leading to dry weather with relatively low humidity. However, other features of the resulting weather are very different in summer and winter (see Figure 7).

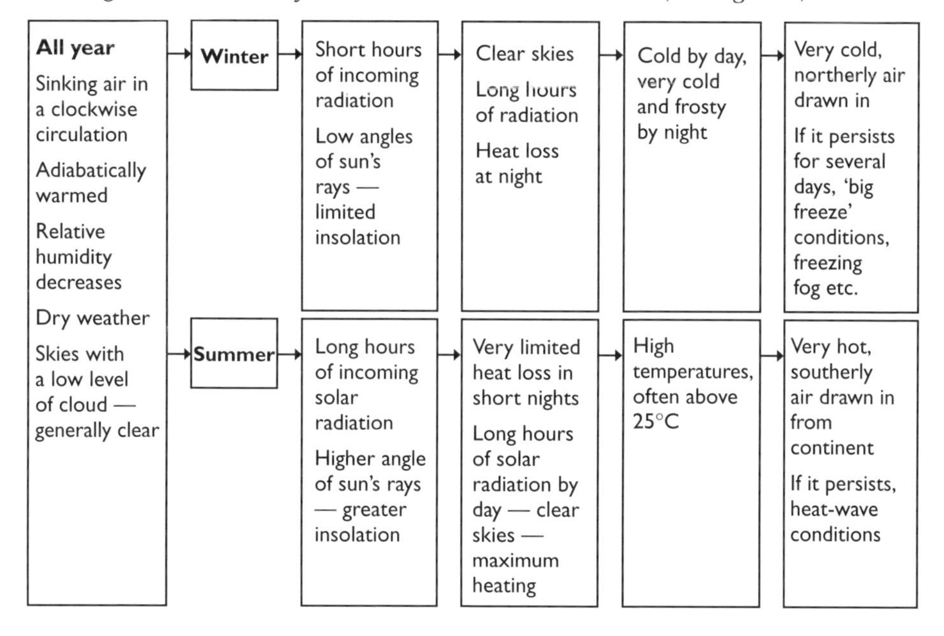

Figure 7 The effects of anticyclones on the UK

Anticyclones are very stable and may persist for many days. They are then known as **blocking** anticyclones because they 'block out' depressions, causing them to track round to the north or south of the large high-pressure systems. Hence, although anticyclonic weather occurs for nearly 10% of the year, it may be the result of only four or five long-lasting systems. The longer the period under the influence of the anticyclonic system, the more extreme the conditions become, and also the more hazardous. Some blocking anticyclones may last for 20 days.

Problems associated with summer anticyclones

- Extreme temperatures can lead to heat stroke and dehydration, with the young and the elderly most vulnerable (e.g. Greece, summer 2000).
- Strong ultraviolet rays can cause skin cancer and cataracts (common in Australia).
- High levels of pollen and fungal spores are trapped near the ground in calm conditions and may trigger hay fever.

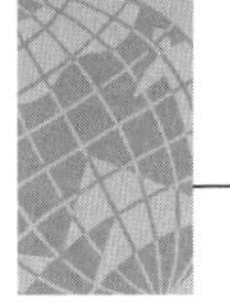

- Forest fires result from spontaneous combustion in the intense heat or from human intervention. They can have a disastrous effect on both wildlife and property (e.g. New South Wales, January 2002).
- Droughts may occur even in countries such as the UK. Excessive evaporation occurs from reservoirs and rivers, and heat-wave temperatures lead to escalating water demand. In the UK, droughts are not expected, so water shortages inevitably result, especially in the densely populated southeast. Drought affects domestic and industrial users and has a major impact on plant and animal life. It poses problems for farmers, in particular those needing to irrigate crops or to maintain quality pasture.
- Photochemical reactions between nitrogen oxides and volatile organic compounds create low-level ozone. Descending air traps the ozone near the ground. It is a problem for those suffering from chest complaints.

Problems associated with winter anticyclones

- Winter smog is associated with temperature inversions and high levels of sulphur dioxide and other gases resulting from burning fossil fuels and traffic congestion. As conditions are calm, pollutants remain in the area. This impacts on health, particularly asthma, lung and heart disease. Some pollutants, such as benzene, can cause cancers.
- Clear skies lead to maximum radiation cooling. Widespread frosts occur, resulting in road accidents and injuries to pedestrians from falls.
- The freezing conditions pose problems for wildlife, particularly birds, which are unable to access food and water.
- The poor and elderly are at risk from hypothermia (hence the fuel allowances and cold-weather payments that are available).
- Freezing fog may occur, which is likely to lead to motorway pile-ups (e.g. the M40 on 28 March 2002).

Table 3 summarises the seasonal weather contrasts for an anticyclone.

Table 3

Winter weather (polar source)	Summer weather (tropical source)
Cold daytime temperatures — usually below freezing to a maximum of 6°C	Hot daytime temperatures — over 23°C
Very cold night temperatures with frosts	Warm night temperatures — may not fall below 15°C
Clear skies by day and night with low relative humidity	Generally clear, sometimes cloudless, skies
Stable conditions so fogs may form, especially radiation fogs in low-lying areas; possibility of freezing fog	Some mist and fogs early morning — especially at the coast where they may be persistent; these are advection fogs created by sea/land differences
As a warm front approaches, the very cold air and ground can lead to heavy snow	Thunderstorms may result from convectional uplift, usually in late afternoon or early evening

Anticyclones and depressions compared

Table 4 compares the characteristics of anticyclones and depressions.

Table 4

	Anticyclones	Depressions
System size	Large — 1800–3200 km across Uniform airmass — Pc or Tc	Small — around 750 km across Contrasting airmasses — Pm and Tm
Pressure pattern	Highest in centre Typically 1024–1040 mb Isobars widely spaced	Lowest in centre Typically 970–990 mb Isobars close together, steep pressure gradient
Circulation	Clockwise pattern of outward-blowing, very light winds of 0–10 knots; calm in centre	Anticlockwise inward-blowing vortex, often strong winds up to 50–60 knots
Stability of system	Stable and slow moving May last from days to weeks	Unstable/mobile system moving rapidly across an area in 24 hours System lasts 2–5 days
Humidity and precipitation	Descending air leads to adiabatic warming; no rain but mist or fog common	Ascending air; cold air wedges underneath warm air, leading to frontal rainfall
Temperature	Hot in summer; cold in winter due to absence of cloud cover	Variety of temperatures as a result of contrasting airmasses

Effects of seasons on climate

Explaining seasonality

Understanding the **migration of the heat equator** is the key to understanding seasonal variation. At the **equinoxes** in March and September, the sun is 'overhead' the equator at midday, whereas at the winter **solstice** in December and the summer solstice in June, the sun is 'overhead' at the Tropics of Capricorn and Cancer respectively (see Figure 8).

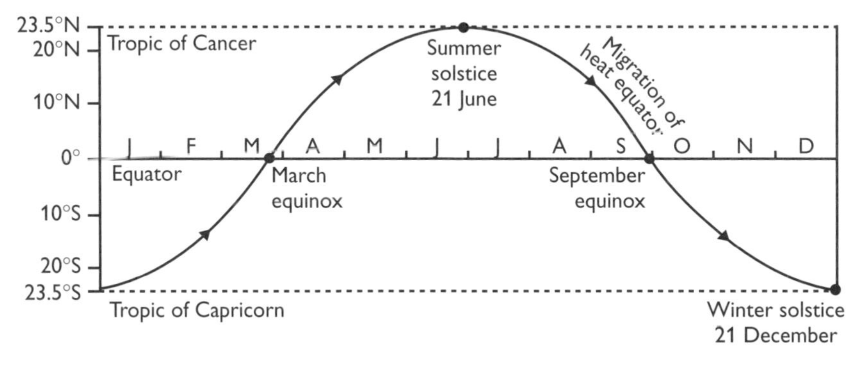

Figure 8 Migration of the heat equator

The concentrated heat causes the air to expand and become lighter than the air north and south. The rising air causes a surface low-pressure zone (the doldrums or equatorial low) which draws airflows towards it, forming the **intertropical convergence zone**, **ITCZ**, or **discontinuity** (see Figure 9).

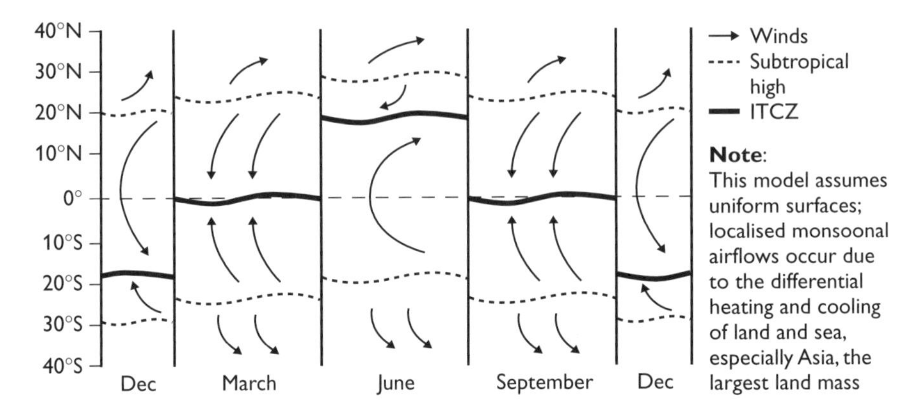

Figure 9 Migration of the ITCZ

The significance of migration of the heat equator is that it triggers changes in the global circulation and its accompanying pressure and airmass belts, which in turn influences the distribution of rainfall (see Figure 10).

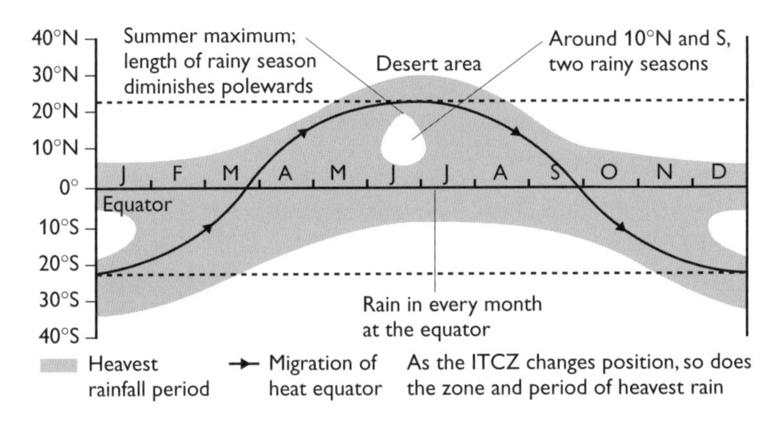

Figure 10 The effects of heat equator migration on the distribution of rainfall

Why do the seasons occur and why does the heat equator migrate?

The Earth is tilted in its orbit around the sun at an angle of 23.5°, so at different parts of its orbit the Earth is tilting in different directions with respect to the sun. On 21 June the sun's rays fall directly over the Tropic of Cancer because the northern hemisphere is leaning directly towards the sun. On 21 December the situation is reversed and the sun is directly over the Tropic of Capricorn. Between these dates, the heat equator tracks north and then south (see Figure 11).

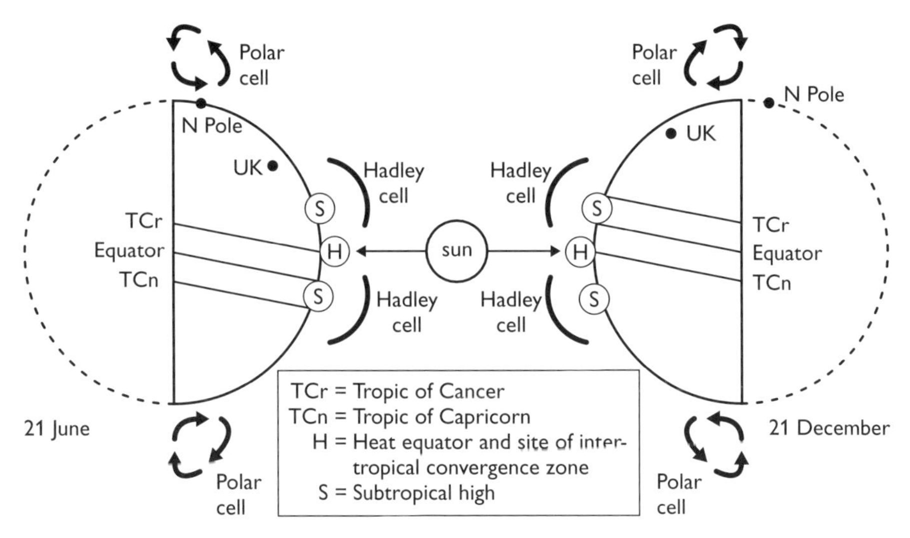

Figure 11 Seasonal movement of the heat equator

What impact does migration of the heat equator have on global circulation?

The atmosphere at the equator receives more concentrated heat, expands and rises. It then spreads out at a higher level in the atmosphere and cools down. This heavier air sinks back to the surface to form subtropical high pressures, which are the source of airmasses such as the maritime equatorial (see Figure 12).

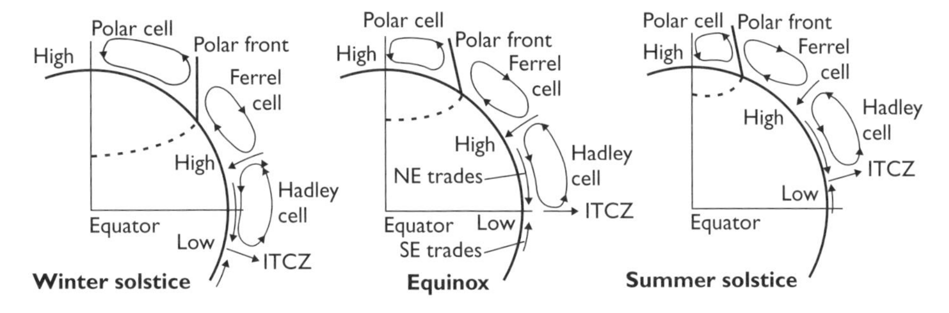

Figure 12 Migration of the heat equator and global circulation

Why is the ITCZ often associated with periods of rainfall?

Tropical air flows from opposite hemispheres to meet at a broad zone (ITCZ). Although there are not the temperature and density differences found in the airmasses converging at the polar front, the warmth and humidity of the converging airmasses leads to potential turbulence and this triggers strong convectional uplift.

What is an ecocline?

An **ecocline** refers to the decline of vegetation as a result of less favourable growth factors such as decreasing temperature, increasing aridity or increasing altitude along a transect. The case study overleaf shows a typical ecocline for a transect from the tropical rainforest to the desert in west Africa.

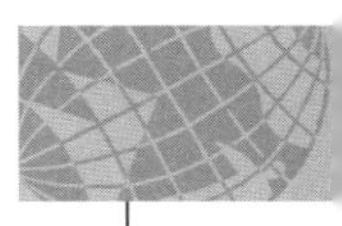

An ecocline: a case study of west Africa

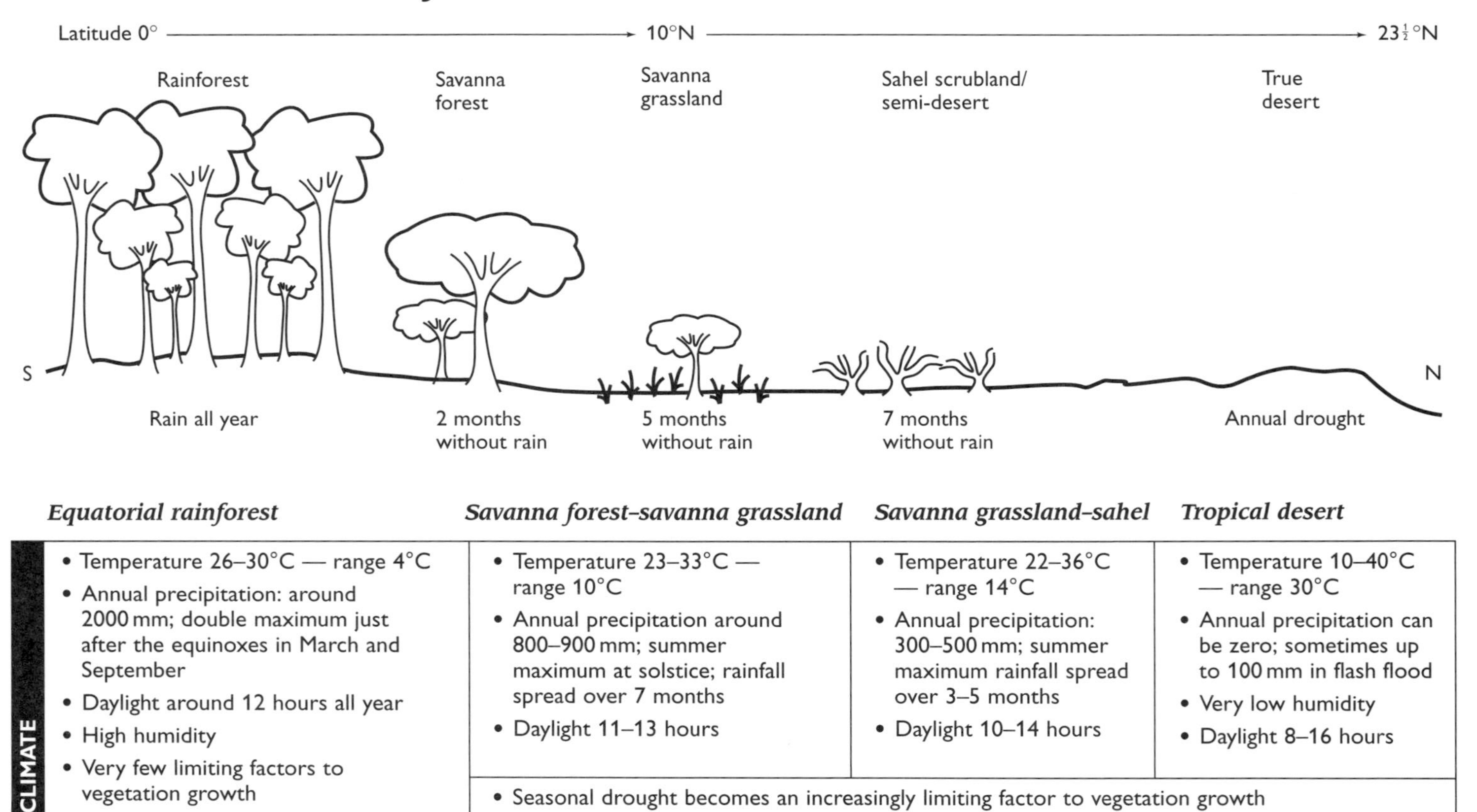

CLIMATE	*Equatorial rainforest*	*Savanna forest–savanna grassland*	*Savanna grassland–sahel*	*Tropical desert*
	• Temperature 26–30°C — range 4°C • Annual precipitation: around 2000 mm; double maximum just after the equinoxes in March and September • Daylight around 12 hours all year • High humidity • Very few limiting factors to vegetation growth	• Temperature 23–33°C — range 10°C • Annual precipitation around 800–900 mm; summer maximum at solstice; rainfall spread over 7 months • Daylight 11–13 hours	• Temperature 22–36°C — range 14°C • Annual precipitation: 300–500 mm; summer maximum rainfall spread over 3–5 months • Daylight 10–14 hours	• Temperature 10–40°C — range 30°C • Annual precipitation can be zero; sometimes up to 100 mm in flash flood • Very low humidity • Daylight 8–16 hours
		• Seasonal drought becomes an increasingly limiting factor to vegetation growth		

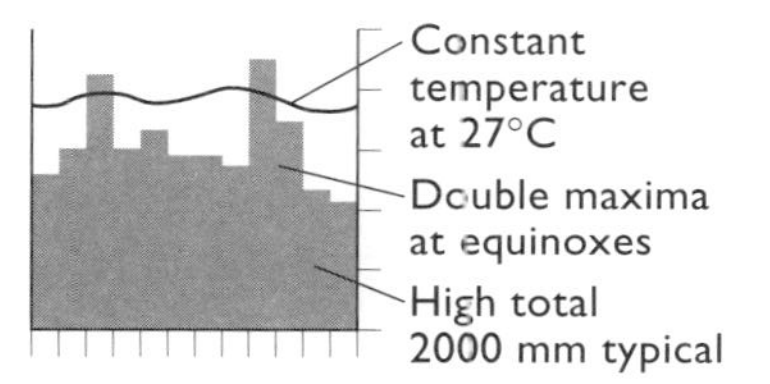

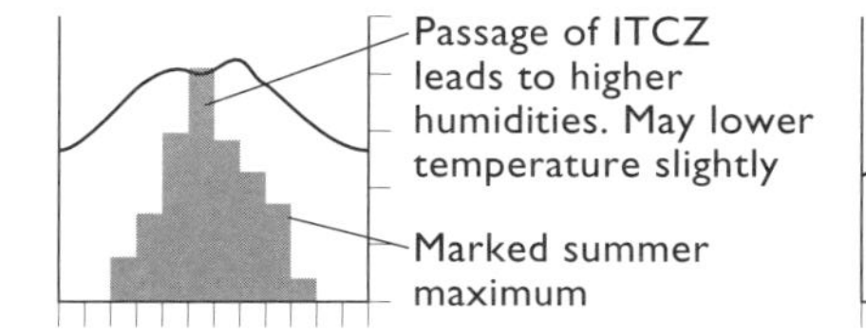

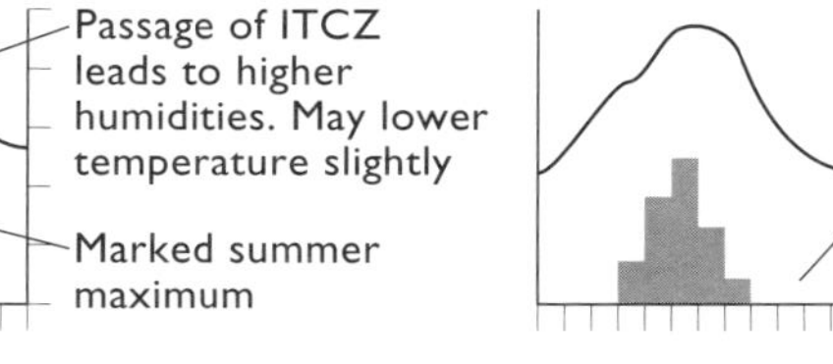

VEGETATION CHARACTERISTICS	• Tall, broad-leaved evergreens in continuous growth • Canopied, stratified forest with a layered structure • High species diversity • Constant heat and humidity lead to high primary productivity with many ecological niches and a large biomass • Rapid nutrient cycling	• Mixed grassland with trees; clearings of woody shrubs and tall trees • Drought-resistant trees such as acacia and euphorbia appear as it starts to get drier (10°N)	• Tall savanna grass; isolated, small, drought-resistant trees; all vegetation develops protection against fire and animal predators	• Xerophytic plants such as cacti, yucca and tamarisk are adapted to arid conditions • Slightly wetter stream course areas support scrub • Limited species
	NPP ($kg/m^2/yr$) 2.2 Biomass (kg/m^2) 45.0 Tropical red earths	NPP ($kg/m^2/yr$) 1.2 Biomass (kg/m^2) 9.0 Lateritic soils	NPP ($kg/m^2/yr$) 0.9 Biomass (kg/m^2) 4.0	NPP ($kg/m^2/yr$) 0.2 Biomass (kg/m^2) 0.6 Desert soils

HUMAN ACTIVITY	• Traditional shifting agriculture in remote areas; otherwise subsistence bush fallowing • Enormous range of tree crops; some plantations such as oil-palms and cocoa • Range of vegetables and staple crops • Forest exploitation	• A mixture of forest and grassland crops, grown according to the wet and dry seasons • Crops include millet, maize, cotton and vegetables	• Largely cattle rearing by tribes following nomadic routes • Severe droughts in recent years; desertification of sahel	• Very limited settlement except at oases • Some mining settlements for oil or precious metals • Nomads herd camels

Short-term climate change

Short-term climate change operates over periods up to 1000 years. At one end of the scale, El Niño is an example of very short-term climatic disruption, whereas at the other extreme, global warming is an example of short-term climate change.

Some people link global warming and El Niño, arguing that global warming triggers the more frequent development of a warm-water cell off the coasts of Peru and Ecuador, which is responsible for disrupting the normal Pacific Ocean circulation and thus causing an El Niño event.

El Niño–La Niña

El Niño is an example of very short-term, widespread **climatic disruption**.

Figure 13 shows that El Niño events happen irregularly and that they are increasing in intensity.

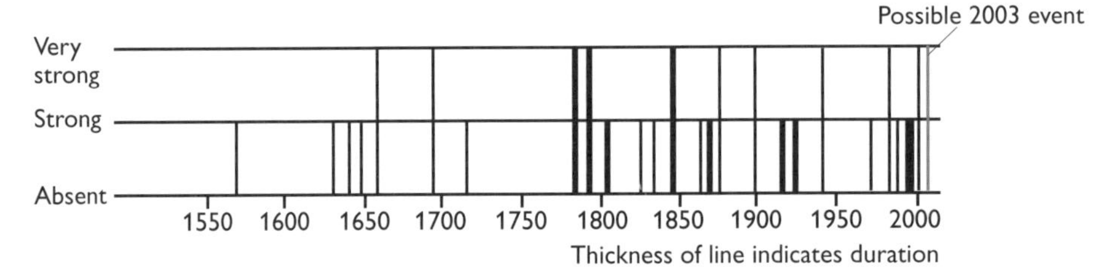

Figure 13 El Niño events

Evidence from biological and physical monitoring suggests that the 2003 El Niño could be beginning. Information from ocean measuring buoys suggests the sea is beginning to warm slightly. This is supported by biological surveys of the feeding habits of birds and fish. A combination of historical records, satellite coverage and computer models should enable scientists to predict the time and magnitude of the event successfully.

When an El Niño event occurs, it affects the climate short-term, not only around the Pacific Basin. It is thought to lead to atmospheric interactions around the world (**teleconnections**), largely because of distortion of the jet stream flows.

There are cold events when the sea surface temperature becomes colder compared with the long-term average for the region. The name **La Niña** refers to the appearance of these colder-than-average conditions. In 1998–99, La Niña caused different disasters in many of the same areas hit by the 1997–98 El Niño. An increased incidence of Caribbean hurricanes was a resultant teleconnection. La Niña usually occurs just before or just after an El Niño event but its occurrence is less frequent.

Prediction of an El Niño or La Niña event does not make them any easier to deal with, because they bring anomalous extreme conditions, often to some of the poorest countries.

Important definitions

- **El Niño** — the appearance of *warm* surface water from time to time in the eastern equatorial Pacific
- **La Niña** — the appearance of *colder-than-average* sea surface temperatures in the central and eastern equatorial Pacific
- **Southern oscillation** — a see-saw of atmospheric pressure between the Pacific and the Indo-Australian area, monitored using the southern oscillation index
- **ENSO** (El Niño–southern oscillation) — the term used to describe the full range of events
- **Teleconnections** — atmospheric connections between widely separated regions
- **Thermocline** — sharp, deep boundary between cold, deep water and the warmer upper layer

How does the El Niño/La Niña cycle occur?

A normal year

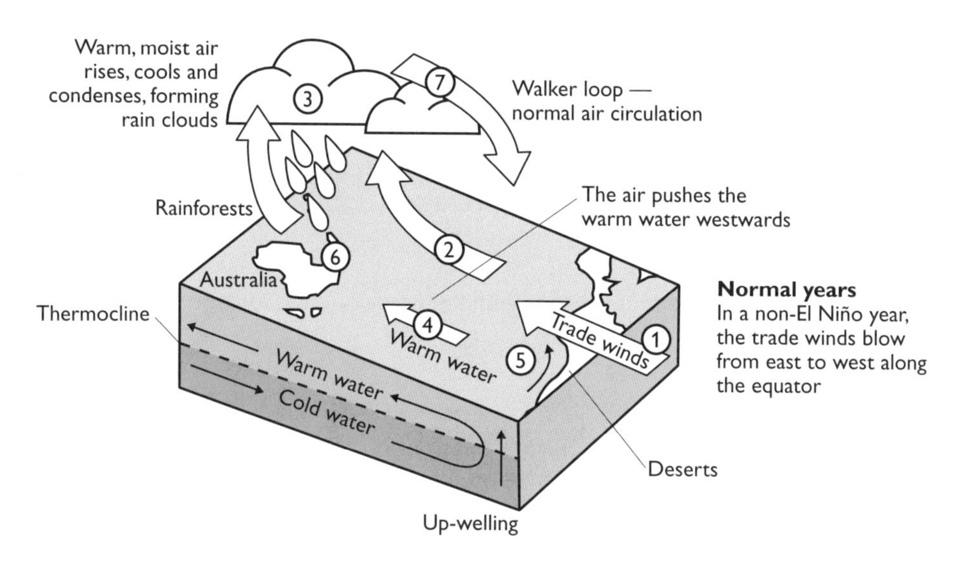

Figure 14

(1) The trade winds blow equatorwards and westwards across the tropical Pacific.
(2) The winds blow towards the warm water of the western Pacific.
(3) Convectional uplift occurs as the water heats the atmosphere.
(4) The trade winds push the warm air westwards. Along the east coast of Peru, the shallow position of the thermocline allows winds to pull up water from below.
(5) This causes up-welling of nutrient-rich cold water, leading to optimum fishing conditions.
(6) The pressure of the trade winds results in sea levels in Australasia being 50 cm higher than Peru and sea temperatures being 8°C higher.
(7) The air sinks, creating the Walker loop.

An El Niño year

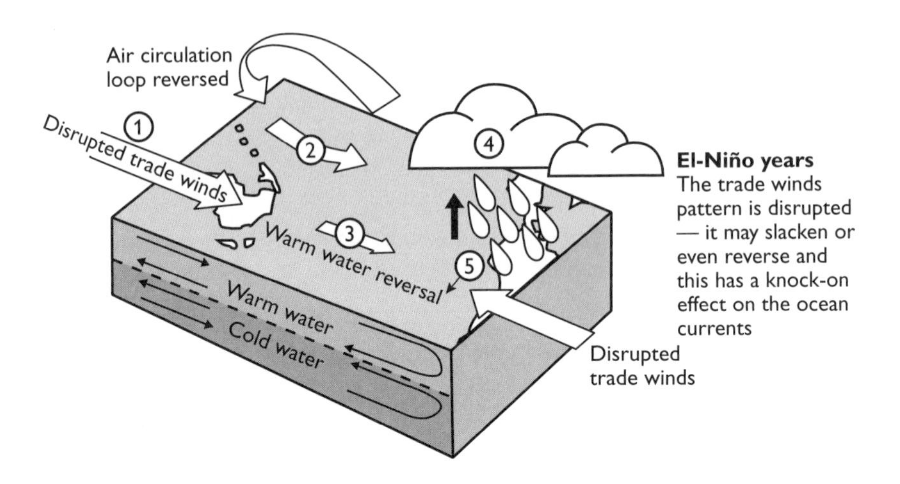

Figure 15

(1) The trade winds in the western Pacific weaken or die.
(2) There may even be a reverse direction of flow.
(3) The piled up water in the west moves back east, leading to a 30 cm rise in sea level in Peru.
(4) The region of rising air moves east with the associated convectional uplift. Upper air disturbances distort the path of jet streams, which can lead to teleconnections all around the world.
(5) The eastern Pacific Ocean becomes 6–8°C warmer. The El Niño effect overrides the cold northbound Humboldt Current, thus breaking the food chain. Lack of phytoplankton results in a reduction in fish numbers, which in turn affects fish-eating birds on the Galapagos Islands.
(6) Conditions are calmer across the whole Pacific.

A La Niña year

This is an exaggerated version of a normal year, with a very strong Walker loop (Figure 16).

(1) *Extremely* strong trade winds.
(2) The trade winds push warm water westwards, giving a sea level up to 1 m *higher* in Indonesia and the Philippines.
(3) Low pressure develops with *very strong* convectional uplift as very warm water heats the atmosphere. This leads to heavy rain in southeast Asia.
(4) Increase in the equatorial undercurrent and *very strong* up-welling of cold water off Peru results in strong high pressure and extreme drought. This can be a major problem in the already semi-arid areas of northern Chile and Peru.

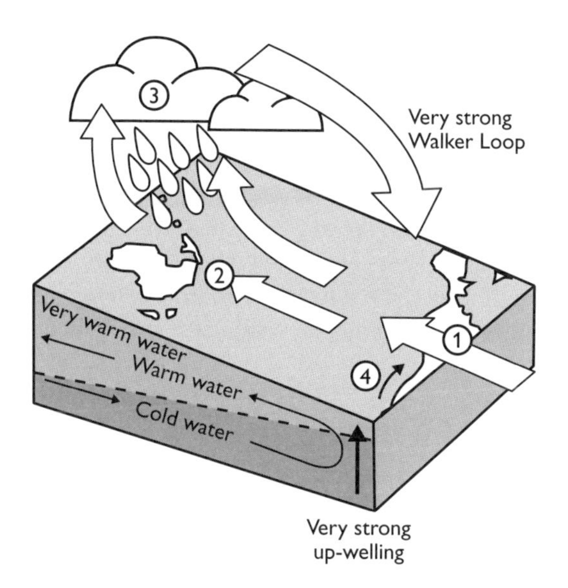

Figure 16

The global impact of El Niño in 1997–98

Direct impacts occurred in the equatorial Pacific. However, researchers are increasingly discovering teleconnections, suggesting that although the impact was piecemeal, the cumulative effects were global.

Most impacts were negative:

- In South America, approximately 1000 people and hundreds of thousands of livestock were killed.
- It caused $20 billion damage.
- Many countries that were badly hit were LEDCs, lacking the capital to purchase technology to cope with the aftermath of the disaster.

There were some positive impacts:

- The Caribbean was almost free from hurricanes in 1997.
- The north Chilean tourist industry flourished as tourists flocked to see the 'desert in bloom'.

The main problem with El Niño is its ability to overturn established climate patterns in a very unpredictable way. The desiccation of Indonesia and eastern Australia led to forest fires in Indonesia and bush fires in Australia. The arrival of torrential rains in western Peru led to flash floods in Ica, accompanied by landslips and followed by a cholera epidemic caused by contaminated water.

Teleconnections occur between the tropical Pacific and the northern hemisphere circulation (the Pacific North America pattern, or PNA). A downstream jet wave appears because of higher than normal pressures over western Canada and lower than normal pressures in southeast USA. In El Niño periods, the jet wave approaching north America is strengthened. The increased warmth in the eastern Pacific leads to increased tropical convection. The moisture pumped into the atmosphere is carried northeast by subtropical jet streams, contributing to heavier rainfall across the southern USA. Other important teleconnections include droughts in Brazil and Africa, which often occur during an El Niño. When the warm pool and the Walker loop move east during an El Niño, the sinking motion is correspondingly displaced, producing drought in the equatorial Atlantic region. The eastward displacement of the Walker loop leads to stronger upper atmosphere winds, which sheer off the tops of developing tropical storms, preventing them from becoming hurricanes in the Caribbean.

Investigating global warming

Global warming is an example of short-term climate change.

Framework for enquiry:

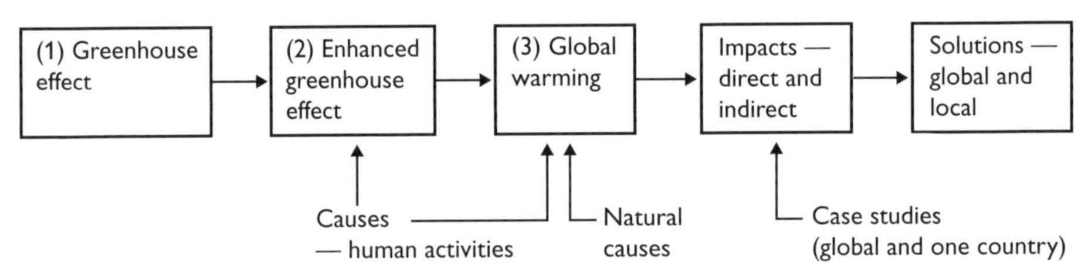

Figure 17

(1) The greenhouse effect

The **greenhouse effect** is a natural phenomenon. It is the process by which water vapour and other gases absorb outgoing long-wave radiation from the Earth and send some of it back to the surface, which is consequently warmed.

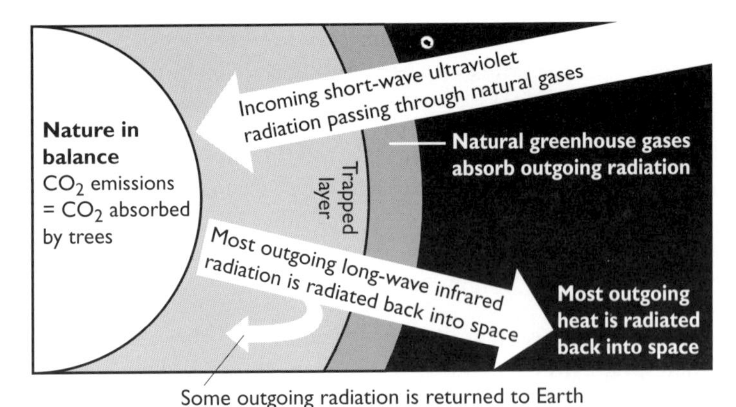

Figure 18

It sustains life on Earth by raising temperatures to a global average of 15°C. The greenhouse gases include water vapour (the biggest contributor), carbon dioxide, methane, CFCs, nitrous oxide and ozone. Until the nineteenth century, the situation was generally stable. Records of carbon dioxide levels taken from core bores up to 8000 years ago suggest that any changes are linked to large, longer-term global temperature swings such as the ice age. Therefore, any large changes in the formation of greenhouse gases must be linked to human activities, which release carbon that has been stored in a variety of sinks.

(2) The enhanced greenhouse effect

The **enhanced greenhouse effect** occurs because of increased amounts of greenhouse gases.

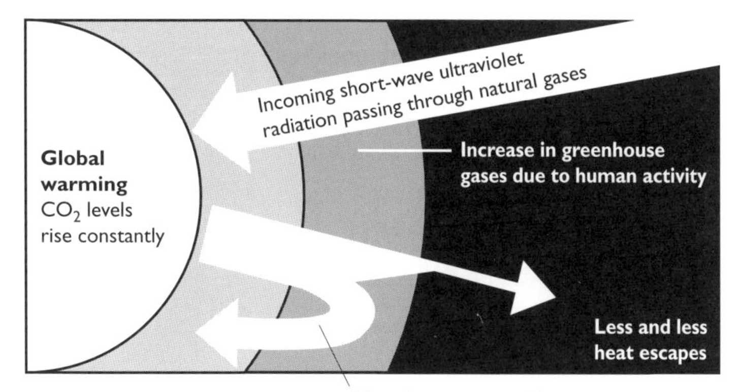

Figure 19

Recent land-use changes, industrialisation and urbanisation have contributed to the increased amounts of greenhouse gases. However, the contribution of greenhouse gases to global warming is not *only* due to increased amounts. Some gases, such as chlorofluorocarbons (CFCs), have enormous global warming potential because their chemical stability means their effects can last for 180 years. The estimated contribution of greenhouse gases to global warming is as follows:

- Carbon dioxide 55–60%
- Methane 15–20%
- CFCs/halocarbons 14–17%
- Nitrous oxide 5–6%
- Others up to 10%

(3) Theories for global warming

A consensus is developing among most scientists that global warming *is* occurring. There is an underlying upward global trend in temperature (Figure 20). The Intergovernmental Panel on Climate Change (IPCC) was set up by the UN to assess the evidence for, and the possible impacts of, global warming. It numbers 2500 leading climate scientists from around the world and is highly influential.

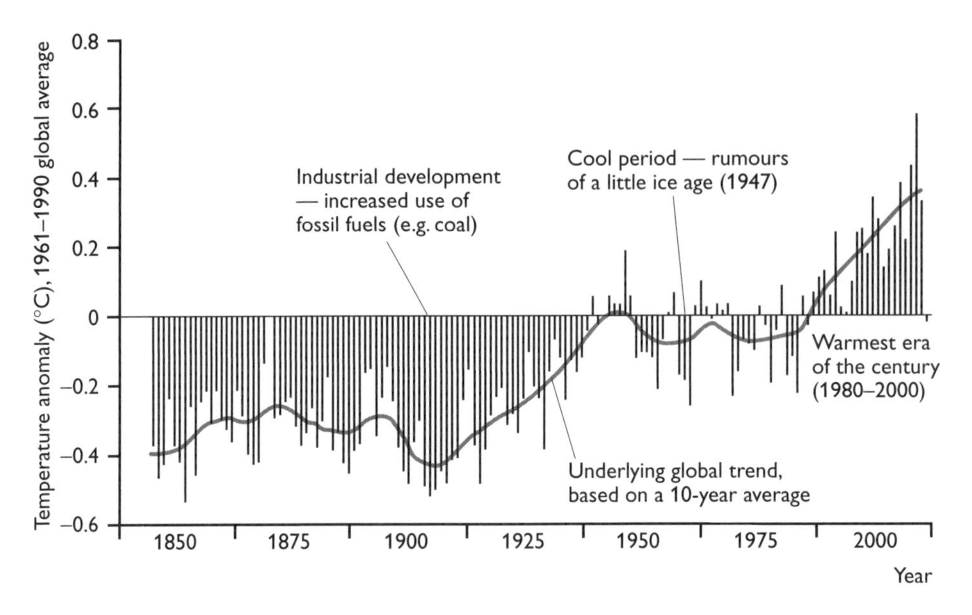

Figure 20 The trend in global warming

The IPCC participates in a global monitoring programme which involves:

- directly observing the atmosphere at both surface and troposphere level using a range of remote sensing equipment
- measuring ocean temperatures
- measuring and monitoring greenhouse gas concentrations
- observing and monitoring sea levels and the extent of glacier and ice sheet melting

Note: rising sea levels are not directly linked to melting of the ice sheets. Thermal expansion as a result of rising ocean temperatures is a major factor.

Evidence for global warming

- Global temperatures have risen up to 1°C over the past 150 years.
- Greenhouse gases have increased since 1850, with rapid rises since 1950, particularly CFCs, so contributing to an enhanced greenhouse effect.
- Sea levels have been rising by an average of 20 cm each century.
- Extreme weather events, anomalous to the established pattern of climate, are becoming more common, perhaps indicating changes in the climatic belts.

Environmentalists are concerned about these changes. They fear that the absence of strategies to control greenhouse gas emissions will result in the situation deteriorating, with serious consequences for food production, water supply, ecosystems and health.

Arguments against the existence of global warming

Some people deny the existence of global warming and cite scientific arguments against it:

- The recent upward trend in temperatures may be a temporary blip and not part of a significant trend. There have been other similar instances in historic times.

- The increase in temperature is small and does not provide grounds for the predictions made.
- Self-regulation will take place. For example, plant biomass increasing with rising carbon dioxide levels would result in increased water vapour, leading to more cloud and producing net cooling.
- Some troposphere measurements suggest cooling temperatures in the upper air.
- The ocean conveyor belt of warm currents, such as the North Atlantic Drift, will be slowed by the melting ice in Greenland, which will reduce salinity (the Spitzbergen Syndrome).

Others accept that global warming may be occurring, but link its occurrence to natural causes:

- Milankovitch cycles lead to variations in the Earth's orbit and tilt, which could lead to climate change.
- Catastrophic volcanic eruptions generate huge quantities of dust, leading to climate change.
- Variations in solar radiation (sun spot activity) could also lead to short-term climatic change.

The consequences and impacts of global warming

Figure 21 summarises the sequence that scientists are modelling in order to assess possible environmental impacts and economic consequences of global warming. They are trying to work out both the scale and the speed of the likely changes. The faster a change occurs, the more difficult it is to adapt to it.

The **eustatic** rise in sea level results from **thermal expansion** of the oceans, which is caused by warming (1–3°C by 2100). It is possible that melting ice sheets in Greenland and Antarctica will also contribute to this. The effect will be global, but its impact will be differential. Worst affected will be some low-lying, heavily populated areas such as Bangladesh and coral atolls such as the Maldives or those in the Pacific. They have had to develop coastal protection strategies, which they can ill-afford.

Other changes are associated with shifts polewards (estimated from 150 km to 550 km in mid-latitudes) of **climatic belts**. This subsequently impacts on the hydrological cycle, ecosystems and economic activities such as farming. The effects are more localised:

- Interior locations such as the temperate grassland areas of the Steppes and the Prairies will be likely *losers*, with even drier summers and even colder winters. This will limit grain yields and make the grasslands even more marginal. As these areas are the world's bread baskets, there is a global issue associated with feeding the world's population, as well as issues of national trade balance and the regional future of farming. The drier summers will place even more demands on scarce irrigation water.
- A likely *winner* will be the Sahel, where the creeping desertification will be arrested and farming prospects will improve because the rainfall pattern will be more reliable as the climate belts shift polewards.

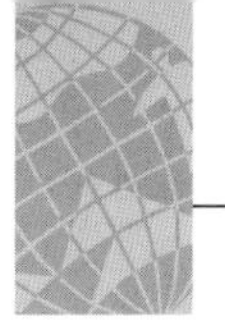

Climate change can lead to increased unpredictability and a higher risk of extreme hazards. It is possible that warming of the oceans could increase the frequency and magnitude of El Niño events. Many scientists predict an increased incidence of storms and hurricanes.

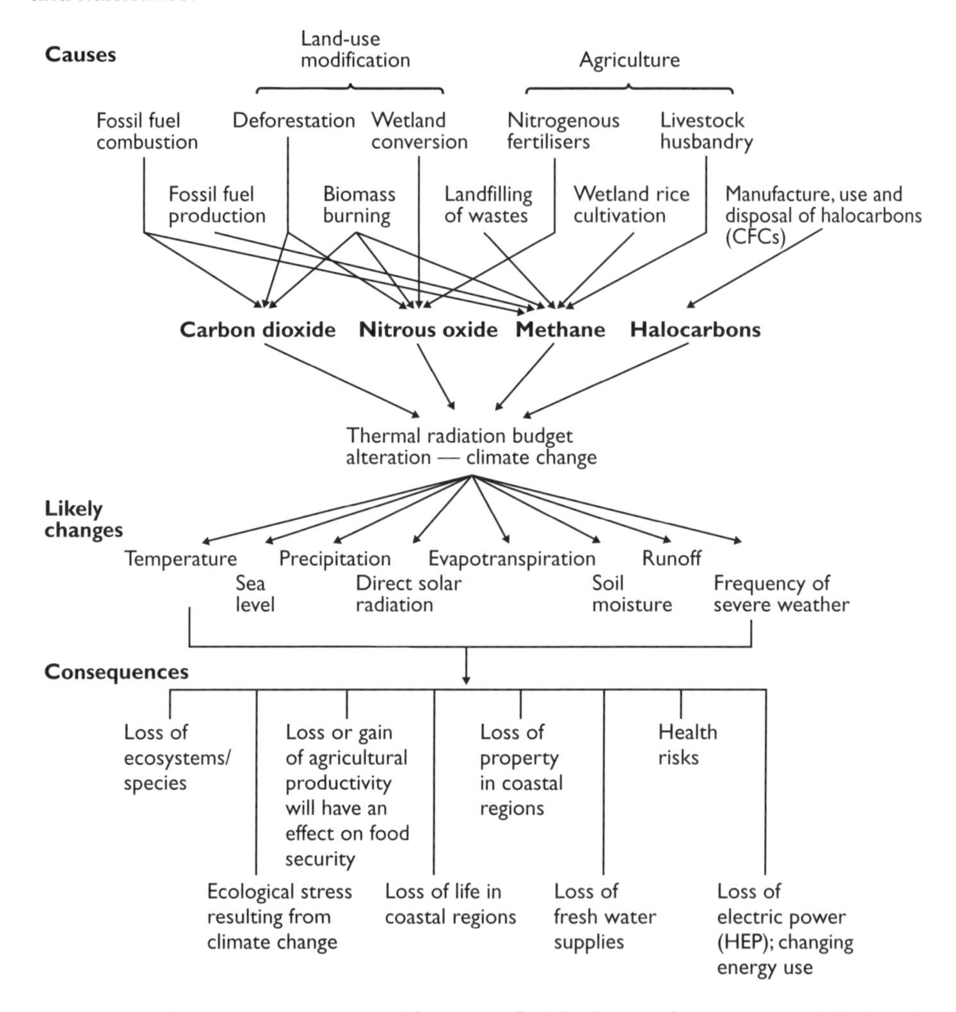

Figure 21 The consequences and impact of global warming — on a global scale

Case study of the UK: what might happen to our climate?

The standard scenario suggests:

- a northwest–southeast divide
- an increased risk of coastal flooding for all low-lying areas, with sea levels up to 50 cm higher by 2050
- autumns and winters will be wetter, especially in the northwest, with more frequent storms
- summers will be drier, especially in the southeast; droughts will be more pronounced

- temperatures in summer will be higher, with twice as many hot days over 25°C; winters will be warmer, without frost in the south
- the climate will become more extreme and more unpredictable
- the climatic zones will move northwards by approximately 300 km, with southern England feeling Mediterranean

There is an alternative scenario, called the Spitzbergen Syndrome. Geographers estimate that the North Atlantic Drift keeps Britain 5°C warmer than expected at our latitudes. Since 1970, there has been a dramatic thinning of the shelf ice in the Arctic and the melting of many arctic glaciers. Arctic ice helps to drive the **Atlantic conveyor**, as the chilly down-flow of a submarine river of cold, dense water helps to draw warm water northwards in its place. The flow of cold, dense water has decreased by 20%. If the trend continues, there will be no warm water from the North Atlantic Drift drawn northwards and Britain will have winters like Spitzbergen, in the Arctic.

What can be done about global warming?

Action to combat global warming can be directed towards:

- modifying the causes
- mitigating and managing the effects

Action towards modifying the causes has to take place at a global level because international agreement is needed to control the emission of greenhouse gases. Global action needs to be supported by national and regional planning and the actions of individuals. Action towards mitigating and managing the effects is necessarily national, regional and local.

The sequence of global action on climate change is summarised in Figure 22.

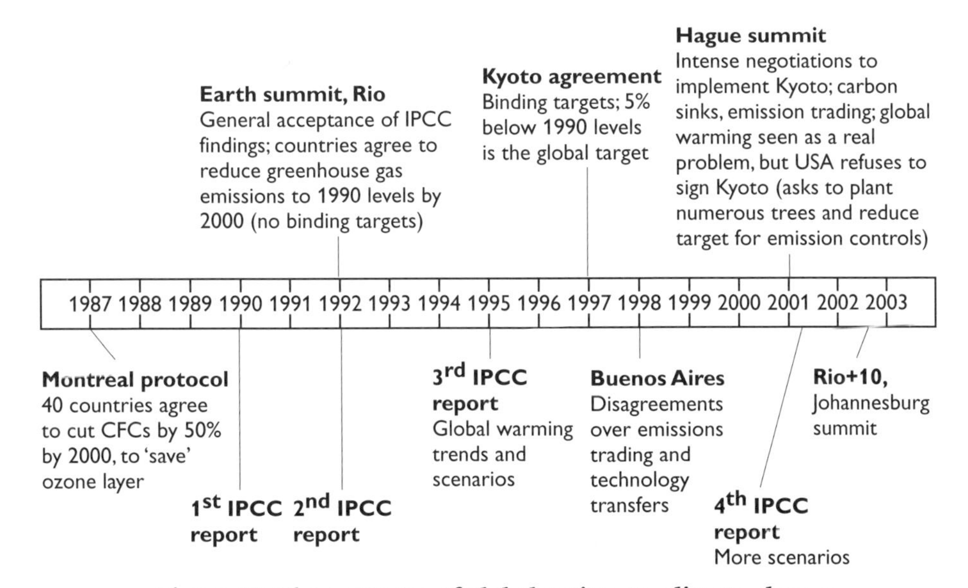

Figure 22 The sequence of global action on climate change

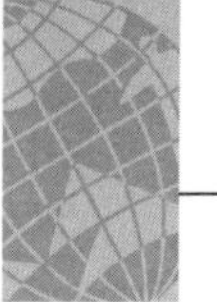

Globally, countries cannot agree on strategies to reduce greenhouse gas emissions. In particular, the USA regards emission control as harmful to its economy, which is based on the heavy use of low-cost fossil fuels. Many LEDCs, such as China, are growing rapidly and want financial help to be available to buy the clean technology needed to cut emissions. Other countries, such as many small island developing states (SIDS), which will be most affected by rising sea levels, just hope action will happen! In 1997, the Kyoto agreement set a legally binding global cut in greenhouse gas emissions of 5.2% below 1990 levels, to be achieved by 2008–12. Developed countries overall have a target to decrease emissions by 5%. Examples of individual targets for selected countries are:

- decrease of 8%: EU and Switzerland
- decrease of 7%: USA
- decrease of 6%: Canada, Hungary, Japan and Poland
- no change: Russia, New Zealand and Ukraine
- increase of 1%: Norway
- increase of 8%: Australia
- increase of 10%: Iceland

Note that all six greenhouse gases are combined.

The 1997 Kyoto protocol developed three mechanisms to help countries reach their targets:

- The development of an **international emissions trading regime**, so that if a country reduces emissions beyond its target, it could sell excess emissions credit to others — **carbon credits**.
- Clean development mechanisms, to enable industrialised countries to finance low-emissions projects in developing countries and to receive credits in return. This includes **clean technology** and planting trees to create extra **carbon sinks**.
- Joint implementation will provide credit for investments in projects in developed and developing countries.

National planning

To achieve their targets, countries such as the UK have developed national strategies for cutting emissions and adapting to the problems likely to be caused by global warming.

Emissions-cutting strategies

- Plans for sustainable energy use and energy-conserving homes schemes
- Strategies for waste, including recycling to avoid methane from landfill
- Move to renewable sources of energy, combined heat and power, cleaner coal
- Green transport strategies — car engines, fuel efficiency, new fuels, car parking, working patterns, public transport
- Air quality regulation — energy and carbon tax proposals
- Agricultural practices programme — feeding strategies for animals
- Planning regulations — compact cities, avoidance of urban sprawl
- Support for emission-cutting technology

Adaptations — the UK climate impacts programme

- Improved water resource management to combat potential summer droughts; management strategies for agriculture; coastal and river flood defence programmes
- Enhanced resilience of buildings and infrastructures, including flood plain zoning and regulation
- Management of wildlife — corridor provision for protected areas to ensure wildlife survival; strengthened biodiversity strategies
- Integrated planning
- Greater raising of awareness among the population
- Improved short- and long-term risk assessment

Introducing the biosphere

Challenges facing the biosphere

Importance of ecosystems

The world's global economy and national economies are based on goods and services derived from ecosystems. Human life depends on the continuing capacity of the biosphere's ecosystems to provide.

Basic goods needed for survival provided by ecosystems include:

- food crops such as fruit and nuts
- food, directly via agro-ecosystems and indirectly through fodder for livestock
- meat and fish
- building materials such as timber
- water
- energy via biomass and hydroelectric power
- genetic resources for medicine

Vital services for survival provided by ecosystems include:

- climate regeneration
- air purification — trees remove carbon dioxide and emit oxygen
- water control — vegetation impacts on the water cycle
- flood protection (forested watersheds) and storm protection (mangroves and reefs)
- water purification — dilutes and carries away waste
- cycling of nutrients
- generation of humus to produce soil
- maintenance of biodiversity and the gene pool
- wildlife habitats
- aesthetic enjoyment and recreation
- employment across all sectors

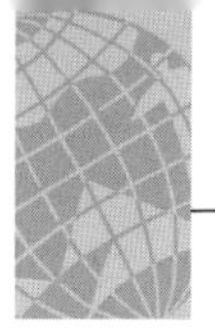

Human impact on ecosystems

If we use up the resources to obtain goods in the short term, valuable services will be lost in the long term. For too long, and at an accelerating pace in both rich and poor nations, development priorities have focused on how much people, businesses and governments can take from ecosystems. Little attention has been paid to the impact of their actions. Recently, global assessments of the condition of the biosphere have been made possible by effective methods of aerial survey and satellite surveillance. Destruction affects *quantity*, and the rate of modification and damaging actions affect *quality*. Together, they impact on biodiversity.

Various indices such as the **living planet index** (Figure 23) have been developed to measure the natural wealth of the Earth's continental and marine ecosystems. The living planet index is the average of three indices, which monitor the changes over time in populations of representative animal species in forests, freshwater and marine ecosystems. It was developed by the Worldwide Fund for Nature (WWF).

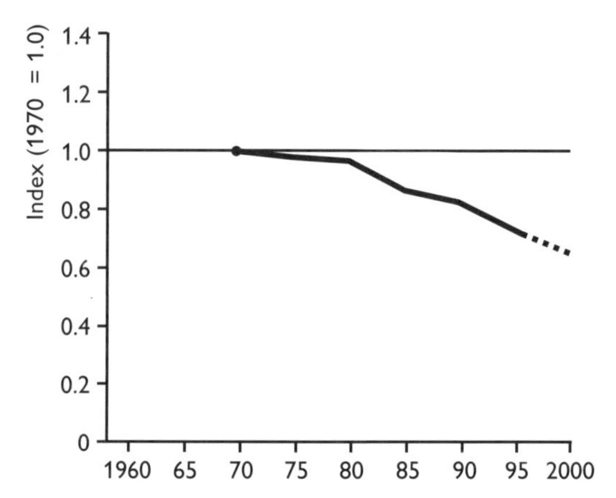

Figure 23 The living planet index, 1970–2000

Between 1970 and 2000, the living planet index fell 33%; the **forest index** used 319 marker species and fell 12%; the **freshwater index** used 194 marker species and fell 50%; the **marine index** used 217 marker species and fell 35%.

Ecological footprint

An **ecological footprint** is a measure of **human pressure** imposed by a given population (on a global, national, regional or local scale) on global ecosystems. It represents the biologically productive area required to:

- produce the food people consume, including imports
- provide resources, such as wood
- provide space for infrastructure and urbanisation
- absorb the carbon dioxide produced by burning fossil fuels
- store and soak up other waste generated

As there is now a global economy, people use resources from all over the world and affect distant places with their pollutants. Therefore, the footprint for each country is calculated to include these areas too.

The world's ecological footprint changes with global population size, the average consumption per person and the resource intensity of the technology used. Changes in the world ecological footprint for the last 40 years are summarised in Figure 24, which shows how it is rising globally.

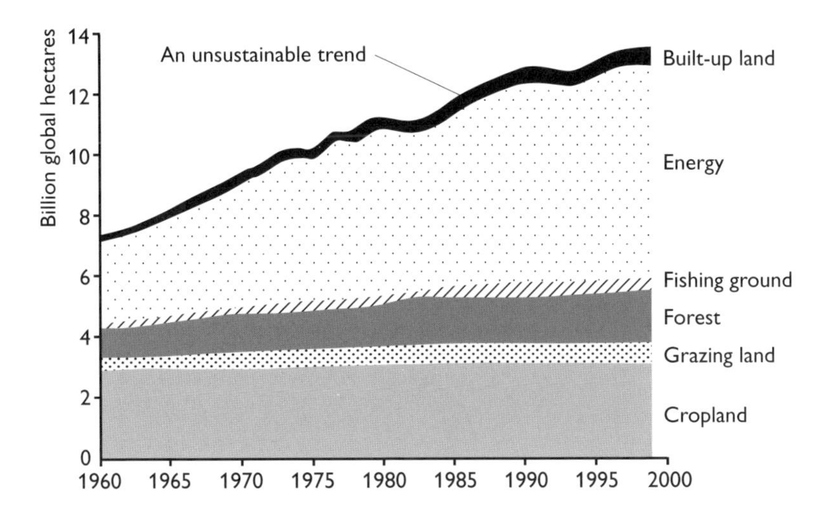

Figure 24 The world ecological footprint, 1961–99 (Source: WWF)

The size of the footprint in world regions is shown in Figure 25.

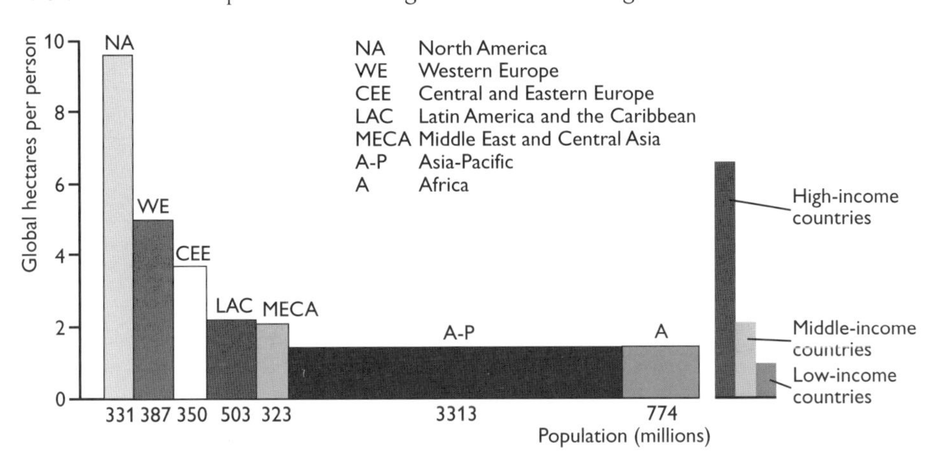

Note: The area of each box is proportional to the footprint of each region (Asia has the biggest). The height of each box is proportional to the region's average annual footprint (North America has the greatest). The width of each box is proportional to the population (Asia has the largest population).

Figure 25 Ecological footprint by region and income group, 1999

The ecological footprint is measured in **area units**. One area unit is equivalent to 1 hectare of biologically productive space with average world productivity. The world average footprint was 2.85 area units, when it was calculated that the available space per person for biological production (**bioproductive capacity**) was 2.0 area units. This means that the area required to produce our food, wood and resources, to build on, and to absorb carbon dioxide emissions and remove waste, is about 30% larger than the area available. This overshoot leads to the depletion of the Earth's natural resources and is reflected in the decline of the living planet index (see Figure 23).

Actions needed to reduce the ecological footprint

These are shown in Figure 26. Essentially, the aim is to reduce the ecological footprint *and* maintain biological capacity.

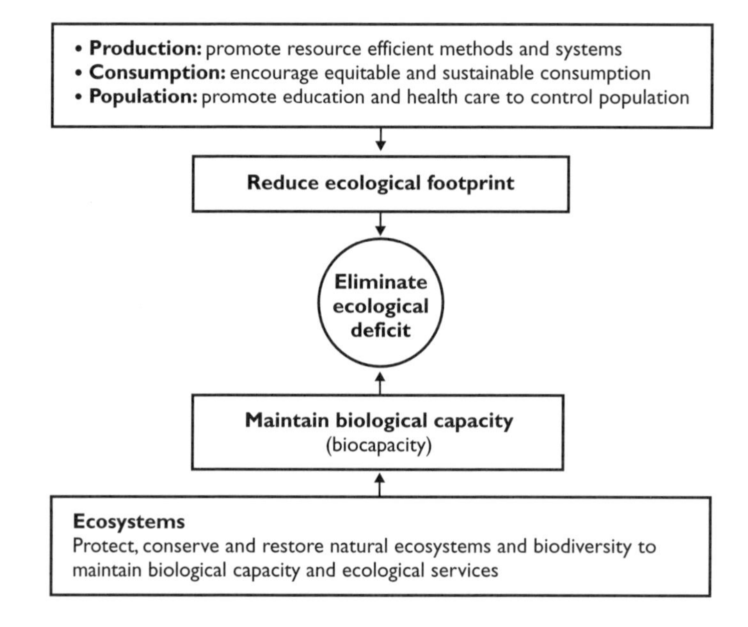

Figure 26 Policies for eliminating the ecological deficit

Understanding global ecosystems

The living world or **biosphere** reaches from the ocean depths and stops in the lower layers of the atmosphere, at the highest point where any living organisms are found. It is the part of the Earth and atmosphere capable of supporting life. The biosphere spans three abiotic environments — the **atmosphere** (air), the **hydrosphere** (water) and the **lithosphere** (earth) — and relies on an exchange of materials between all three. The **Gaia hypothesis** suggests that the biosphere regulates the environment, maintaining optimal conditions for survival.

Ecosystems

Ecosystems are defined as self-regulating, biological communities in which living things (the biotic component) interact with the environment (the abiotic component). Ecosystems vary in scale:

- biosphere — the global ecosystem
- biomes — world-scale communities of plants and animals characterised by a particular vegetation type, such as tropical rainforest
- regional scale — such as peat bogs
- micro-scale — such as lichens on a rock or a single tree

Figure 27 is a model of an ecosystem, with inputs, outputs and stores.

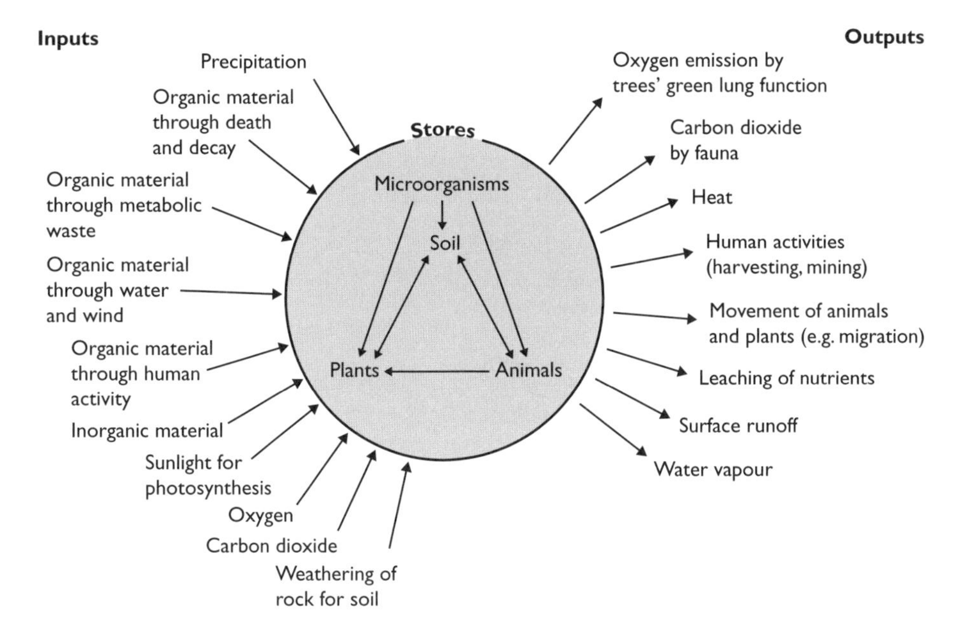

Figure 27 A generalised ecosystem

Ecosystems have two basic functions — to redistribute energy and to recycle nutrients.

Energy flow

Energy cannot be created or destroyed, but only transferred from one state to another. Energy from the sun is fixed by plants during photosynthesis and may:

- be stored as chemical energy in plants or animals — the **biomass**
- be passed through the ecosystem feeding levels (**trophic levels**) through **food chains** and **food webs**
- escape from the system as outputs of material — decay, excreta or heat energy from respiration

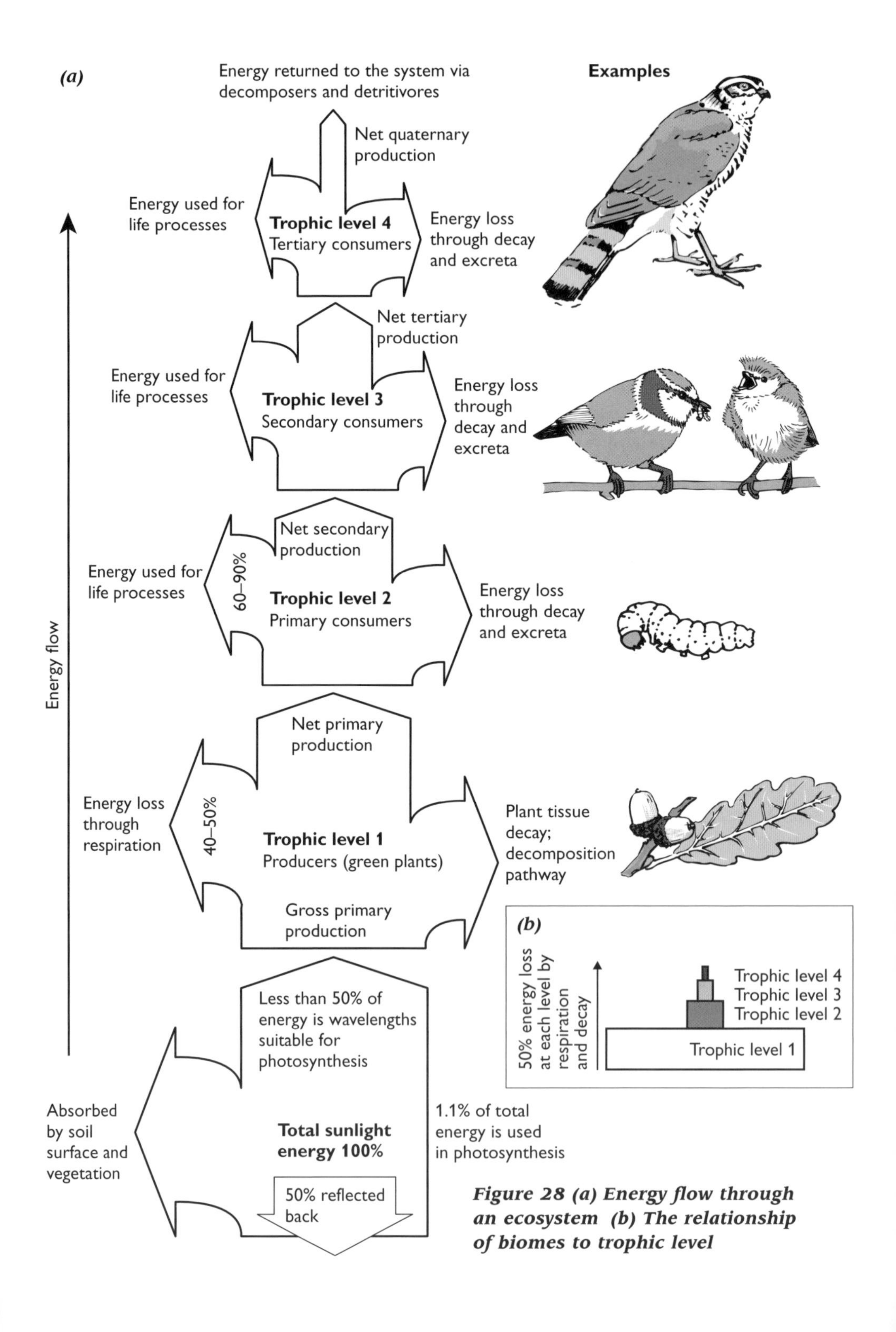

Figure 28 (a) Energy flow through an ecosystem (b) The relationship of biomes to trophic level

Optimum conditions of heat, light and humidity lead to large numbers of primary producers, which support greater biodiversity in a more complex food web with more ecological niches.

Primary productivity is the rate at which energy is converted into organic matter. It can be measured by the amount of new **biomass** produced each year.

Ecological productivity depends on:
- heat (temperature), which controls the rate of chemical reactions
- water, which is a key component in many chemical reactions
- nutrient availability
- light for photosynthesis

Maximum primary productivity occurs where these factors are optimum, for instance in equatorial areas or shallow tropical waters.

High primary productivity supports high **biodiversity**. If any one of these four factors is in short supply, it **limits** overall **primary productivity**.

Solar radiation is the key controlling factor, via insolation. The amount of solar radiation reaching an area varies with latitude, altitude, seasonality and day length.

Gross primary productivity (GPP) is a measure of all the photosynthesis that occurs within an ecosystem, whereas **net primary productivity** (NPP) is the energy fixed in photosynthesis minus the energy lost through respiration (R).

$$NPP = GPP - R$$

NPP is the new growth available for other levels of the food chain to use. It is measured as a dry biomass. NPP and plant biomass vary for different types of global terrestrial and marine ecosystems.

Nutrient cycling

Nutrient cycling occurs within an ecosystem. Nutrients are the chemical elements and compounds organisms need to grow and function. Nutrient cycling and energy flow are interdependent:
- The rate of nutrient cycling may limit the rate that energy is trapped. Plants cannot grow and trap more energy if essential nutrients are absent.
- Energy flow may also limit the rate of nutrient cycling. If a plant is not capturing enough energy for life processes, then its death will result in breakdown of its organic matter.

Global ecosystems and management of threats

Note: The three sections that follow are options. You need to study only one. It is vital that you can:

- compare the main types
- analyse their distribution
- outline their ecological value
- understand the threats facing them
- discuss solutions for their management

1 Forest biomes

Figure 29 shows the current distribution of forest biomes.

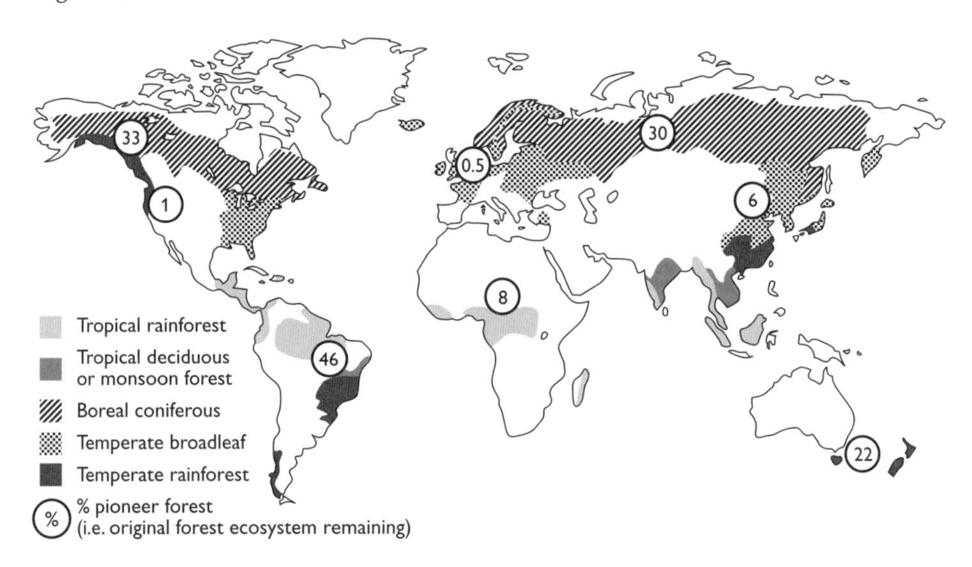

Figure 29 The worldwide distribution of forest biomes

Forests used to cover around 65% of the Earth's land surface. As recently as 1950 the forested area was 45%. By 2000 it was just 25%. Various modification processes have resulted in only 38% of this remaining cover being undisturbed, original or pristine forest. Often, as forest areas are exploited, any replacement forest consists of fast-growing non-native species. These semi-natural plantations support lower levels of biodiversity.

Two factors explain these trends:

- escalating demands for timber — for building, fuelwood, wood products and paper
- accelerating demands for forest land — for agriculture, mining and urbanisation.

This can lead to fragmentation, which accelerates the impact of hunting and poaching and makes the forest areas more vulnerable to pests and disease

Temperate deciduous forests in western Europe and eastern America were the first to disappear, followed by tropical forests in southeast Asia. Today, the valuable hardwood rainforests are under threat. There is conflict between indigenous forest dwellers and commercial companies.

Physical factors, such as rainfall, originally limited forest cover. A minimum of 400 mm of *effective* precipitation is required for tree growth. Physical factors also influence the types of tree found, with markedly different distributions of evergreen, deciduous and coniferous trees. However, it is human exploitation of forests that influences both the present-day distribution of cover and biodiversity levels (**bioquality**). Forests could be the **climatic climax** of many more areas.

Tropical rainforest profile

Location and distribution

Tropical rainforest is found mainly between 10°N and 10°S of the equator, in the Amazon basin, the Congo basin, coastal western Africa, Malaysia and Indonesia. A deciduous variant was originally found in India and Madagascar. Regional circulation of airmasses extends rainforests in some places to the tropics.

Climate

True rainforest is associated with an equatorial climate — a narrow, annual temperature range of 27–31°C and annual rainfall between 1800 and 3000 mm with no month having less than 100 mm. Rainfall tends to be heaviest just after migration of the heat equator. Constant heat and humidity aid development of evergreen forest. Tropical deciduous forest can adapt to a wider temperature range of 18–32°C and up to four drier months.

Soils

Soils are largely ferrallitic, tropical red earths. They are surprisingly infertile because nutrients decay rapidly, and run off or leach away. Most nutrients are stored in the biomass — hence the disastrous effect of deforestation.

Natural species

Tropical rainforests are home to more than 50% of the world's animal and plant species. Species differ in each rainforest province. Trees include mahogany, greenheart, palm, and rubber and numerous rare hardwoods. Trees rarely grow in stands.

Plant adaptation

- Evergreen trees photosynthesise all year.
- Buttress roots support very tall trees.
- Drip tips to leaves shed rainfall.
- Large stomata promote transpiration.
- Trees are adapted to compete for light.
- Adaptation of epiphytes — plants which live on trees, but are not parasitic.

Structure

Rainforest has well-defined stratification. The emergent layer can be up to 50 m; the canopy is at 40 m and the sub-canopy at 20 m. The floor layer is dark with little undergrowth. Epiphytes, lianas and saprophytes are common. Biodiversity is high, particularly of birds and primates living in the canopy zone. Nutrient cycling is rapid.

Major issues

- Numerous products, such as fruit, rubber, vanilla, rattan (palms), palm oil, nuts and cocoa need to be harvested annually.
- Logging for tropical hardwoods is responsible for 40% of rainforest destruction.
- Development of roads and services by logging companies has paved the way for other developments, such as cattle ranching and mineral exploitation in Brazil.
- Rainforests are home to 1.5 million indigenous people, who have a unique knowledge of the diversity of the forests and medicinal properties of plants. Some practise shifting cultivation over a wide area and require large supplies of fuelwood.
- Deforestation has impacts such as loss of biodiversity, hydrological change, climate change and soil destruction. This loss of services is a major global concern.

Temperate deciduous forest profile

Location and distribution

Deciduous forests usually occur from 45° to 60°, N and S. They formerly covered large parts of Europe, North America, northeast Asia, parts of Tasmania and New Zealand. They have largely been cleared for agriculture.

Climate

They are associated with a cool, temperate, maritime climate — summer temperatures of 15–20°C, winter temperatures of 4–6°C and rainfall of 600 to 1800 mm annually. The trees need a frost-free period of 8 months.

Soils

Soils are fertile brown earths, with **mull humus.** Nutrient cycling is good, with litter incorporated into the humus.

Natural species

Broad-leaf trees predominate — oak, ash, elm and beech, with maple and hickory in North America. There are frequently several dominants in an area.

Plant adaptation

- Deciduous leaves drop before low winter temperatures occur, reducing transpiration. In winter, primary productivity decreases because there is less light, temperatures are lower, soil moisture is less easily absorbed and there are no leaves for photosynthesis.
- Broad, thin leaves absorb maximum sunlight in summer.
- Forest-floor plants flower rapidly in spring when light reaches the floor. They survive the winter as underground storage organs, such as bulbs or tubers.

Structure

Deciduous forests are clearly stratified into trees, shrubs, herbaceous plants and ground layers. They alter with the seasons and support a wide variety of animals, some of which hibernate or migrate in winter. Herbivores include rabbits, mice and squirrels; carnivores include foxes, owls and weasels. The food web is complex and biodiversity is high.

Major issues

- Very little natural woodland remains. Most is managed as part of estates. Woodland has been used to supply high-quality timber, willow and wood for charcoal.
- The woodland is used for stock grazing, for example sheep, but this is damaging.
- Increasing use for game management — fox hunting, shooting (farm diversification grants).
- Aesthetic qualities and recreational use — walking, riding and mountain-biking trails.
- The woodland provides habitats for wildlife — species can be conserved by management practices, such as coppicing.
- Quality soils have resulted in most woodland being deforested and the land given over to agriculture. Recent afforestation, while beneficial with broad-leaved trees, does not have the biodiversity of the original woodland.

Boreal forest profile

Location and distribution

Boreal forests occur north of latitude 60°N in Scandinavia, Siberia, Canada and Alaska. Apart from high altitudes, such as the Andes, they are absent from the southern hemisphere. This is because there are no large landmasses at appropriate latitudes.

Climate

They are associated with a cold continental climate — summer temperatures of about 18°C, winters as low as –25°C, with 6 months below 0°C. Strong winds and the high wind-chill factor result in moisture being rapidly evaporated or frozen. Precipitation is 400–500 mm per year with a slight summer maximum and winter snowfalls.

Soils

Podzols, developed from acid, leached soils, are poor quality with a deep litter layer, **mor humus** and poor nutrient supplies.

Natural species

Four types of conifer predominate — spruce, pine, fir and larch. Further south (55°N) there is more variety as deciduous species occur. Beyond the Arctic circle, boreal forest declines into tundra.

Plant adaptation

- Evergreen trees start photosynthesising as soon as the temperature is high enough.
- Waxy needles reduce transpiration.
- A pyramid tree shape means that branches can withstand heavy snow and strong winds.

- The trees are shallow rooted, so they can take moisture from the active layer of permafrost as soon as it thaws.
- Thick, resinous bark protects trunks from extreme cold.
- Cones protect seeds.
- The trees are adapted to soils with poor supplies of nutrients.

Structure

Stratification is simple. Limited light inhibits understorey shrubs (cranberry), mosses and lichens. Low productivity and the harsh climate limit the animals. Herbivores include caribou, elk and reindeer; carnivores include wolves and bears.

Major issues

- Conifers are grown in stands. They are important worldwide as a source of softwood and paper.
- Commercial felling is frequently clear-cutting, but reafforestation takes place with stands of secondary plantations. Only 5% of forests are original ancient or natural forest.
- Intensive forest management leads to a loss of biodiversity, particularly rare bird species.
- The need to earn foreign hard currency has led to heavy logging in Russia, with many accessible areas heavily harvested and the land used for agriculture.
- Tree death caused by acid rain can be a major problem.

Nutrient cycling in forest biomes

Tropical rainforest

Tropical rainforest has a large above-ground biomass (B) in which most nutrients are stored, leading to high biodiversity. Leaf fall occurs all year. Decay is rapid, so the litter nutrient store (L) is small. The soil store (S) is small because of rapid leaching due to high precipitation, which also causes loss by runoff. Also, soils are very deep, often 10 m from the parent material. There is active nutrient circulation. Problems occur with deforestation as the soil is starved of nutrients.

Figure 30

Tropical rainforest

Precipitation

B

L

S

Runoff

Leaching

Weathering

Deciduous forest

Deciduous forest has a relatively large above-ground biomass store with annual leaf fall in autumn. Mull humus (pH 6–7) leads to rapid decay. Incorporation of litter into the soil means that the soil nutrient store is large. Leaching is less significant than in rainforests. Under favourable conditions, nutrient cycling is rapid, with little loss from the system. The more equal nutrient distribution makes this system less fragile.

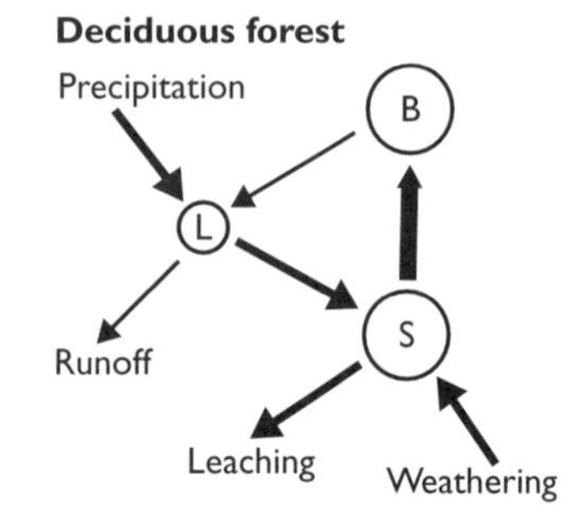

Boreal forest

The biomass in boreal forests acts as a small store of nutrients. Cold conditions lead to smaller trees, adapted to strongly acidic soils. The litter layer is the largest store of nutrients, as the mor humus leads to acid conditions that inhibit breakdown and decay. The soil is a poor nutrient store because the acid conditions lead to podzolisation, an extreme form of leaching. This system has a poor nutrient cycle, with slow replenishment.

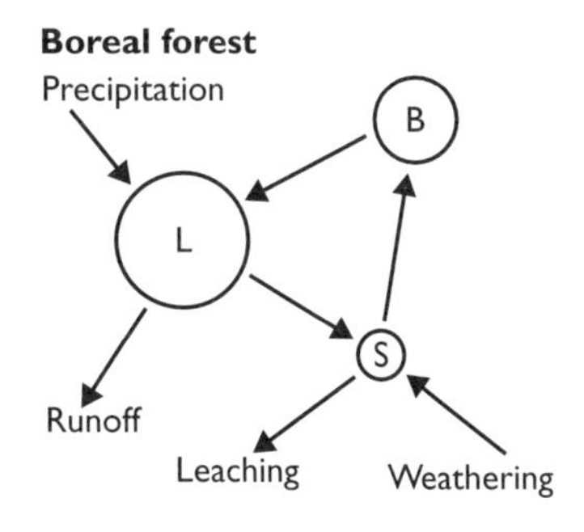

Primary productivity in forest biomes

From boreal to tropical rainforest, the complexity of food webs, the availability of ecological niches, biodiversity and the range of animals all increase.

Table 5 compares the forest biomes in terms of net primary productivity and biomass.

Table 5

	Net primary productivity per unit area ($g/m^2/year$)	Mean primary productivity ($g/m^2/year$)	Biomass (kg/m^2)	Mean biomass (kg/m^2)
Boreal forest	400–2000	800	6–40	20
Temperate deciduous forest	600–2500	1200	6–60	30
Tropical rainforest	1000–3500	2200	6–80	45

The combination of rapid nutrient cycling and high primary productivity in rainforests compared with other forests adds to the value of goods and services provided. In contrast, boreal forests have low primary productivity and a more open, poorer quality nutrient cycle. In spite of lower levels of biodiversity and a more limited range of goods and services, their softwood for paper and building is valuable. The complexity of the food web and the variety of ecological niches is directly related to primary productivity levels.

Threats to forests

Temperate and tropical forests cover approximately equal areas of the Earth's surface. There is currently a greater threat to tropical and original temperate rainforests, because of the quality of the hardwood. Deforestation is the worst threat. Tropical forests tend to be more fragile, because of the importance of biomass as a nutrient store. Also, because of the high biodiversity, many ecological niches are specialised and, if lost, affect the food chain.

In general, the amount of temperate forest has slightly increased in recent years. However, often new afforestation is coniferous and lacks the quality of the original forest cover. Present-day policy is to replace forests after cutting down and to develop new areas.

Most tropical forests occur within LEDCs, so the threats are different and include threats to indigenous peoples and their way of life, through commercial exploitation, mainly by transnational companies.

The future

The solutions can be applied to forests in general, but reflect the differences between tropical and temperate forests. They are concerned with sustainable use of forests, maintenance of biodiversity and achieving optimal environmental and social impact.

Key ideas include:

- Protection and conservation of forest reserves — the best to be in virgin forest of high bioquality. Reserves linked by natural corridors to guard against threats such as climate change.
- Use of buffer zones around reserves.
- Conservation and maintenance of forests on highly sensitive watershed areas.
- Fencing off key forest areas, such as natural or ancient woodlands, against animal grazing.
- Development of agro-forestry — cropping beneath tree crops in tropical forests.
- Marketing and certification of sustainable timber products by ITTO, and the Forestry Stewardship Council (FSC). Also debt swapping, in return for LEDCs conserving forest areas.
- Avoidance of clear felling, maintenance of permanent logging roads, leaving trees to allow post-forest recovery — all aim to control post-logging soil erosion.
- Enforceable, locally agreed working plans for exploitation of forests such as felling cycles and management of forestry operations to sustain yields.
- Development of multi-purpose forestry including recreation, controlled hunting and zoning to avoid conflict.
- Empowering local people to develop extractive reserves; new initiatives such as beekeeping or butterfly farming; bio-prospecting to develop products from medicinal plants; value-added wood products.

2 Grassland biomes

The potential loss of services provided by grasslands has received much less attention than high-level campaigns to save rainforests and reefs. However, grasslands are arguably more important to a much larger percentage of people. Grasslands are home to 1 billion people — nearly 20% of the world's population. Moreover, many grasslands are drylands. People subsist on these marginal lands, with scant, variable, unreliable rainfall. This makes these areas particularly susceptible to damage by human activities. They are slow to recover from degradation resulting from mis-management, such as overgrazing caused by the need to increase food supplies. The **desertification** of many semi-arid grasslands is a major global environmental problem. Many tropical grassland areas are also areas of potential famine.

Grasslands are important because they:

- are a source of cereals, meat, milk, wool and leather
- provide habitats for a range of wildlife
- constitute a vital genetic resource for crop breeding
- are a source of potential energy and fuelwood
- conserve carbon and regulate water in the same way as forests
- are important areas for recreation and tourism

Grasslands contribute to biodiversity. They constitute 19% of the world's centres of plant diversity, 15% of the endemic bird areas and 30% of the WWF's unique eco-regions.

It is estimated that up to 40% of the Earth's land area could have been covered by **natural** grasslands. Grassland has been reduced by grazing modifications, conversion to agriculture, mining (e.g. open-cast coal mining in Wyoming) and urbanisation. Up to 90% of the tall grass prairies of North America have been destroyed.

The locations of tropical and temperate grasslands, together with estimates of the amount of natural grassland remaining for some important areas, are illustrated in Figure 31.

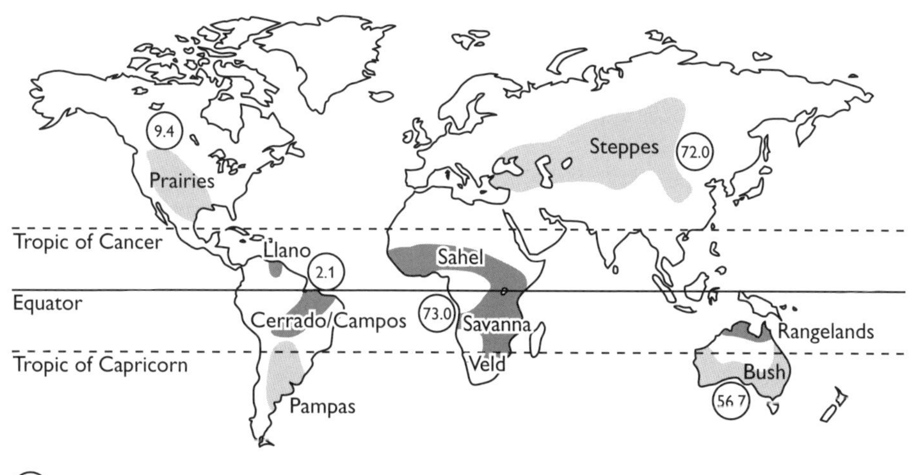

Figure 31 The worldwide distribution of grassland biomes

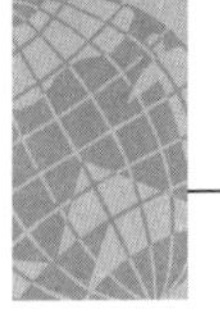

Grasslands are largely devoid of trees in both temperate and tropical areas. This is the result of human actions, such as fire, grazing or use of fuelwood, since in many areas conditions are wet enough for trees to survive. The primary factors influencing grasslands are climatic. Grasslands survive in areas of low rainfall, or where heat and strong winds lead to high evaporation rates. Grasses have a rapid life cycle, so they can grow immediately after periods of frost and drought. Their dense root network enables them to cope with unpredictable rainfall. Locally, impoverished soils may lead to grasslands, for example savannas found on laterite soils in areas wet enough to support trees. In other areas, grasslands occur because of human-related factors such as grazing or destruction of trees by fire. These grasslands are **plagioclimax** vegetation, *not* **climatic climax**.

Tropical grassland profile

Tropical grasslands occur approximately within latitudes 5–20°N and S in central Africa, Brazil, Venezuela and northern Australia. In Africa these grasslands are noted for their wide variety of game, such as antelope and wildebeest.

Climate

The climate type is tropical, wet–dry, with temperatures of 22–33°C and annual precipitation of 300–1500 mm, concentrated in summer. The growing season varies from 3 to 8 months. It is limited by the lack of rainfall in winter — the dry season.

Soils

Soils are largely ferruginous, tropical red soils, including latosols. Leaching is incomplete and upward capillary action concentrates iron, which can lead to the formation of laterites.

Land use

- Traditional herding (e.g. by Masai)
- Wild game parks
- Cash crops, such as peanuts and cotton

Plant adaptations

Drought-resistant xerophytes, such as thorn scrub and acacia, grow in the driest savanna (Sahel). Some plants, such as the baobab tree, develop corky, fire-resistant bark.

Main issues

- Threats to nomadism — overpopulation pressurises grazing, fuelwood and water supplies in Africa.
- Recurrent fires have led to a loss of biodiversity.
- Challenges of cash cropping, for example cotton and coffee.
- Game reserve management — illegal hunting and poaching.
- Increasing unpredictability of rainfall — desertification.

Temperate grasslands

Temperate grasslands are widely distributed between latitudes 40° and 55°N and S. Native animal species include bison and buffalo.

Climate

The climate is temperate continental, with temperatures of 15–20°C and annual precipitation of 250–600 mm (late spring–summer maximum), falling as snow in winter. The growing season averages 5 months and is limited by low winter temperatures.

Soils

Chernozems (black earths) occur in long grass prairie; chestnut soils occur in short grass prairie. Leaching occurs after spring, owing to snowmelt. Capillary action causes calcium accumulation. The soils are deep and productive.

Land use

- Traditional herding and hunting, for instance in the Mongolian Steppes but no longer in North America
- Ranching and intensive arable farming of cereals or soya beans

Plant adaptations

The plants are nearly all grasses. Wind chill from strong winds and physiological drought limits tree growth. The high ratio of leaf surface to ground area ensures sufficient productivity through photosynthesis in markedly seasonal climates. Many grasses have rhizomes (underground stores for maximum absorption of water etc.).

Main issues

- Loss of biodiversity — introduced species for improved pastures.
- Under threat from agriculture, especially in North America.
- High-technology farming leads to environmental damage, for example eutrophication.
- Extreme fragmentation — few natural areas left, especially in the USA.
- Excessive farming of marginal lands — soil erosion in steppes and prairies (dust bowls).

Structure and functioning of grasslands

Nutrient cycling in tropical grasslands is more rapid than in temperate grasslands. In tropical grasslands the ferruginous soils contain few nutrients. This is because decomposition is slow in the dry season and leaching occurs in the wet season.

In temperate grasslands, chernozem soils are a rich store of nutrients, returned to them from the litter by bacteria. The climate is relatively dry, so nutrients are not leached away.

The savannas are on average 50% more productive than the prairies, because the higher temperatures and rainfall result in a larger biomass (see Table 6).

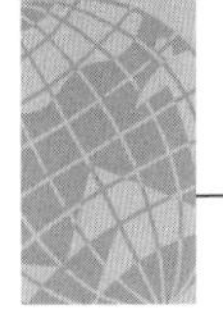

Table 6

Biome	Net primary productivity (NPP) (kg/m²/year)	Biomass (kg/m²)	Ratio of biomass:NPP	Percentage of the world NPP
Temperate grassland	0.6	1.6	2.7	5.4
Tropical grassland	0.9	4.0	4.4	13.5

The higher NPP impacts on food chains and biodiversity. Savannas are valuable wildlife habitats. They have large numbers and a wide range of primary consumers (herbivores, such as antelopes) and of secondary consumers (carnivores, such as lions).

Protecting the biodiversity of grasslands

Measures include:

- establishing **protected areas**, not just national parks for 'big game' (enclosing these animals actually attracts poachers and hunters). Buffer zones and conserved areas where activities such as ecotourism are allowed work well.
- saving or conserving **keystone species** whose survival has the greatest influence on the functioning of whole ecosystems, such as mychorrizal fungi or kangaroo rats whose digging of burrows helps the establishment of shrub seedlings. Projects have tended to concentrate on high-profile, media-catching campaigns or on vulnerable large mammals, which have less impact on the health of the whole ecosystem.
- controlling **grazing pressure**, for example by limiting water supplies in certain areas or fencing to rotate grazing.
- **habitat restoration**. This can be costly and slow but some measures can help:
  - recreation of natural hydrological conditions
  - reintroduction of lost species, for example from existing conservation areas
  - fencing against herbivores such as cattle
  - introduction of livestock grazing systems
  - planting to control erosion
  - fertilisation of pastures to promote growth
- developing **local initiatives** to conserve biodiversity using skills of the community, for example the CAMPFIRE project in Zimbabwe.
- introducing **soil conservation techniques**, such as the Canadian National Green Plan.
- establishing more **research projects** to monitor biodiversity levels and develop new indicators. Grasslands have hitherto received less attention than reefs or rainforest.
- establishing **regional** action plans to cover grasslands that cross international boundaries.

3 Marine ecosystems

There is a wide range of marine environments (Figure 32) which support ecosystems with very different levels of primary productivity and biodiversity (Table 7). Oceans cover 71% of the Earth's surface.

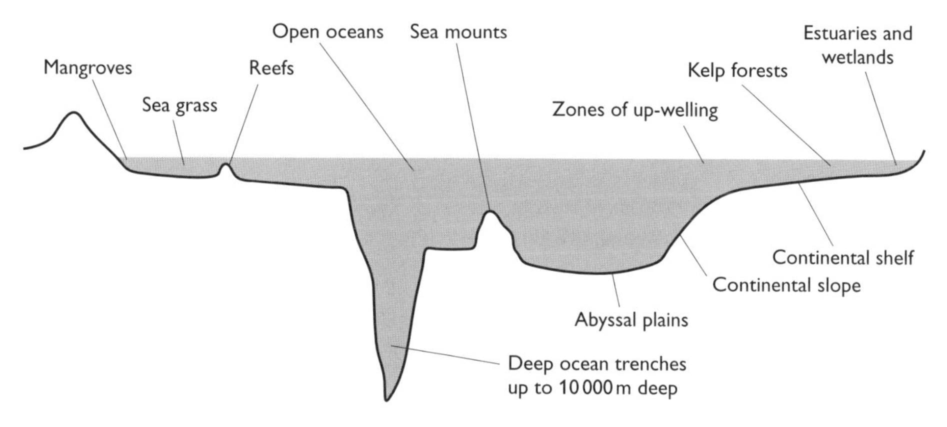

Figure 32 The range of marine environments

Table 7

	Mangroves	Corals	Open oceans	Up-welling zones	Continental shelves	Estuaries
Primary productivity* (g/m²/year)	200–2500 (1500)	500–4000 (2500)	2–400 (125)	400–1000 (500)	200–600 (360)	200–3500 (1500)
Biomass† (kg/m²)	1.0	1.2	0.003	0.02	0.01	1.0
Biodiversity‡	Highest	High	Variable	Dependent on system	Quite high	Highest

* Their very large area means that the oceans are the largest contributor to productivity.

† Systems for gaining information on species and monitoring them are less well developed than on land, especially in the ocean deeps.

‡ Overall productivity is determined by the nutrient supply, which comes from many sources including runoff from the land, decaying sea life and the atmosphere.

Pressure on mangroves and corals, from land-based activity such as siltation and pollution, is high. There is less pressure on the open oceans because of their inaccessibility. However, exploitation of rare species is an issue, for example the seahorse trade. The up-welling zones are concentrated fishing grounds because of the large amounts of plankton available for fish (e.g. anchovies) to feed on. Continental shelf areas are major fishing grounds — overfishing is an important issue. The introduction of alien species to up-welling zones and continental shelf areas can have a disastrous impact on food chains. Estuaries suffer land-based pressures, usually for

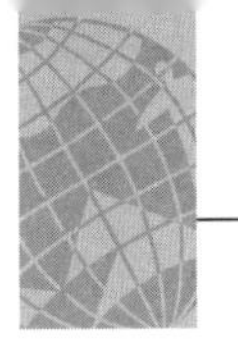

development. There are issues of siltation and pollution from urbanisation and industrialisation. The continental margins are under greatest pressure because they are accessible to both direct exploitation and coastal development. Figure 33 shows the distribution of these threatened reefs and mangroves.

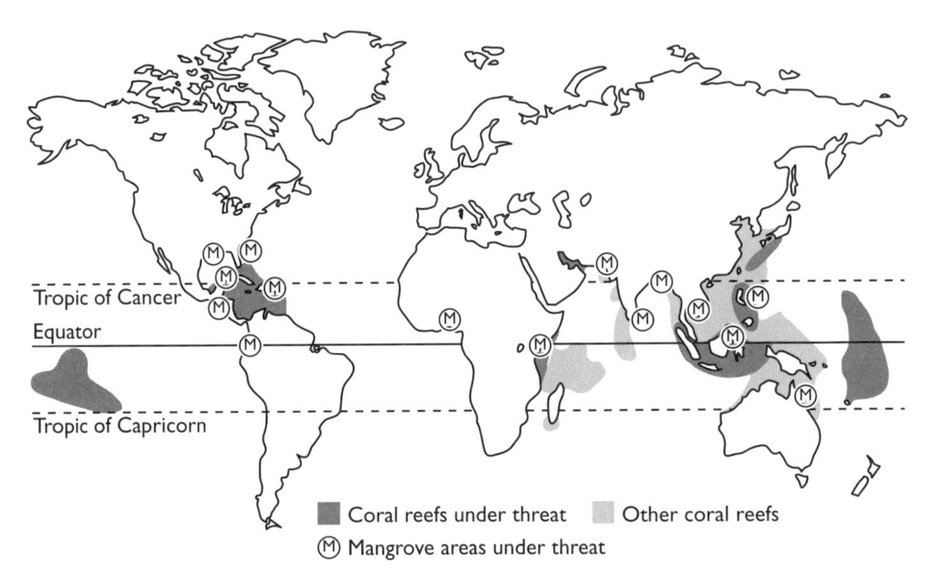

Figure 33 The distribution of coral reefs and mangroves that are under threat

Marine ecosystems, because of their variety and their contrast to land-based ecosystems, provide a portfolio of vital, often irreplaceable, goods and services (Table 8).

Table 8

Goods	Services
• Fish and shellfish constitute about 16% of the protein consumed worldwide — 1 billion people depend on fish as their primary protein source • Fishmeal — animal food, fertilisers • Seaweed — food and in pharmaceuticals • Construction materials — sand, gravel, coral • Salt • Water — in arid areas, after desalination • Supply of genetic resources	• Guard against storm impacts (e.g. mangroves) • Provide habitat for wildlife (e.g. as a breeding area for fisheries) • Maintain biodiversity — few species have yet been discovered • Dilute and treat wastes as well as serving as a dump • Provide harbours and transportation routes • Provide human habitats — 40% of the Earth's peoples live within 100 km of the coast • Provide employment — fishing, tourism and manufacturing • Provide aesthetic enjoyment and recreation • Act as a major carbon sink and oxygen source because of the high productivity of phytoplankton

Corals, mangroves and sea grass in tropical areas, and the coastal and shelf areas in temperate zones, are under the greatest pressure from biodiversity loss. However, the undisturbed ocean deeps are affected by dumping of waste and pollution from land-based activities. The Global Biodiversity Strategy in 1996 identified six fundamental causes of marine biodiversity loss:

- The high rate of population growth and consumption of natural resources, especially in LEDC coastal areas.
- The steadily narrowing spectrum of traded products taken from oceans, because of, for example, an obsession with cod and haddock.
- Economic systems and policies that fail to value the environment and resources of the oceans.
- Inequality in the ownership, management and flow of benefits from the use and conservation of biological resources of oceans.
- Deficiencies in knowledge of marine ecosystems and in monitoring procedures.
- Legal and institutional systems that promote or permit unsustainable exploitation.

The damage caused to northern ocean food webs is now spreading to remote southern oceans, so it is now a global problem.

Marine biodiversity loss

Factors contributing to marine biodiversity loss include:

- habitat loss and fragmentation in marine 'hot spots' such as coral reefs
- changes in water flow as a result of deforestation, abstraction for irrigation and HEP dams, which can affect coastal wetlands
- excessive or inappropriate exploitation of resources, for example dredging for sand and gravel
- genetic improvement and selection in aquaculture, leading to loss of natural species
- over-fishing by large factory ships at the expense of indigenous peoples
- use of netting systems which damage key species such as dolphins and tuna
- pollution from industry, mining, aquaculture, fishing, fish processing, farming and urbanisation, which affects all habitats, especially coastal, and has some long-term food chain impact
- changes in water sediment loads, for example from watershed destruction or hard engineering schemes
- climate change — short-term impact of El Niño and other oscillations; global warming, which leads to coral bleaching and has a potential impact of flooding from rising sea levels; changes in salinity levels
- introduction of exotic species from aquaculture, the aquarium trade and ships' ballast tanks, for example tropical jellyfish decimated salmon in the Hebrides in 2002

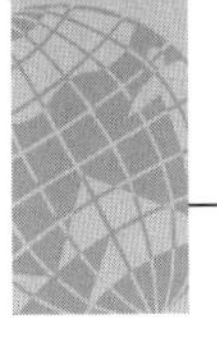

Profile of corals

Distribution

Reef corals are found between latitudes 30°N and 30°S in marine water with a minimum temperature of 18°C. Temperatures above 33°C lead to bleaching and disease. The water is less than 30 m deep, allowing light through for photosynthesis, and is clear from siltation. Zooxanthellae (algae) survive in these conditions. They provide oxygen needed by the corals.

Importance

Coral reefs are important in themselves. Also, because of very high primary productivity by the algae, reef environments support a wide diversity of marine life (many ecological niches).

Structure and functioning

- Coral reefs have high biodiversity. They have complex food webs and are the source of food for millions. They also provide subsistence living for many small island coral atoll communities.
- They provide sheltered lagoons and protect against coastal storms.
- They are spawning and nursery areas for reef fish, providing 25% of the fish/shellfish harvest and 90% of animal protein in many small island communities.

Threats

Direct threats which cause outright destruction include coral mining and blasting for fish. Indirect threats from land-based activity also result in degradation. The main causes are dredging, reef-based tourism, pollution and siltation. Natural threats include bleaching as a result of global warming, disease and hurricane damage.

Profile of mangroves

Distribution

Mangroves grow in the marginal tidal zones between tropical sea and land, in swampy muddy waters. They prefer brackish water and are salt tolerant. They are a partner system to coral reefs which are found between the same latitudes.

Structure and functioning

Mangroves are adapted to tidal areas. They conserve water via thick waxy leaves, which reduce transpiration during low tides. They may have **stilt** roots, covered with breathing cells, which draw in air. Their anchor roots support them in mudflats.

Mangroves:

- provide a sheltered environment for fish to breed
- trap sediment, keeping ocean waters clear
- protect against storm waters and winds
- provide fuelwood and building materials
- trap pollutants such as heavy metals and dioxins and reduce these to harmless compounds
- provide numerous medicines

Threats

The main threat is from land-based development, which completely destroys the mangroves. In individual countries up to 85% have been lost, mostly in the last 50 years. Globally, 50% have been lost, although there is now some replanting.

Mangrove forests are used for building and fuelwood and suffer from land-based and water pollution.

Solutions to marine issues

As Figure 34 shows, solutions need to combine actions:

- across a variety of scales from local to global
- across a spectrum of management from conservation and totally exclusive protection, usually of a single species, through to sustainable development, which allows local people to manage their resources and derive economic benefit from them

	Protection	**Protection with limited eco-developers**	**Protection with sustainable economic development**
Local	*In situ* creation of marine reserves	Use of buffer zones around marine reserves	Agenda 21 frameworks; bottom-up management
Regional	Some *ex situ* conservation, e.g. seahorse nurseries Species management via national quota system	ICR framework for action — capacity building; research and monitoring	↕ Marine Management Areas (e.g. SMMA, St Lucia) New methods of cultivation, e.g. rice–fish culture; fish farms; diversification projects (e.g. INTERFISH, Bangladesh)
National	Large marine ecosystems		
International global	Convention on International Trade in Endangered Species (CITES); world heritage sites; IUCN framework of protected sites	Conservation of biological diversity International Coral Reef initiative	UN law of the sea establishes codes of conduct etc. for responsible fishing

Figure 34 Solutions to marine issues

Global ecofutures

What is biodiversity?

Biodiversity is the variability among living organisms in terrestrial, marine and other aquatic ecosystems and the ecological complexes of which they are part. It can be divided into:

- species diversity
- genetic diversity
- ecosystem (habitat) diversity

A further consideration is the rarity of the species (endemism) or their habitat.

Does biodiversity matter?

Species diversity influences ecosystem stability. A variety of plant and animal species is needed so that ecosystem functions, such as carbon or hydrological cycling, occur with maximum efficiency. Diversity bolsters an ecosystem's resilience, that is, its ability to withstand climatic change, such as long-term drought or global warming.

Genetic diversity can determine an agro-ecosystem's resistance to pests and diseases, thus influencing its productivity — hence the concern about plant breeding for agribusinesses, which tends to reduce naturally diverse polycultures to artificial monocultures, such as rice or wheat. The **gene pool** is fundamental to human health and to the combating of disease. The loss of knowledge from threatened indigenous peoples and the physical loss of key species threatens traditional medicine and the search for drug ingredients.

Ecosystem (habitat) diversity involves the complete food web — the biotic and abiotic components which represent a visual resource, for example in reefs and rainforests.

Factors influencing biodiversity

While the general concerns are about loss, especially that induced by humans, there are positive factors that increase biodiversity, as well as negative ones that threaten it.

> *Hint:* Many questions ask you to assess the impact of human activities on biodiversity. To get a balanced view, you need to look not only at the immediate and underlying causes of ecosystem damage but also at conservation strategies.

Positive factors encouraging biodiversity

- High temperature, light and humidity levels, together with a lack of seasonality, encourage high primary productivity, which supports a diverse, complex food web.
- Complex structures in ecosystems provide more ecological niches.
- Large continuous biomes support a bigger range of plants and animals. Extensive boundaries encourage migration.

- A greater range of altitude provides more diverse habitats.
- Isolation on islands encourages endemism (uniqueness) of species — they develop in a distinctive way (e.g. Galapagos or Madagascar).

Negative factors which threaten biodiversity

- Natural factors include any limiting factors to growth — a dry or cold environment reduces both the total numbers and the range of species (remember trophic pyramids).
- Small size and isolation increases vulnerability to disaster — corridors may be needed for species to migrate away from changes in temperature and rainfall.
- Human factors vary from place to place but can be divided into two categories — immediate and underlying (root) causes.

This means that biodiversity varies from place to place and threats to biodiversity also vary, usually from country to country. The model shown in Figure 35 can be applied to your own case studies.

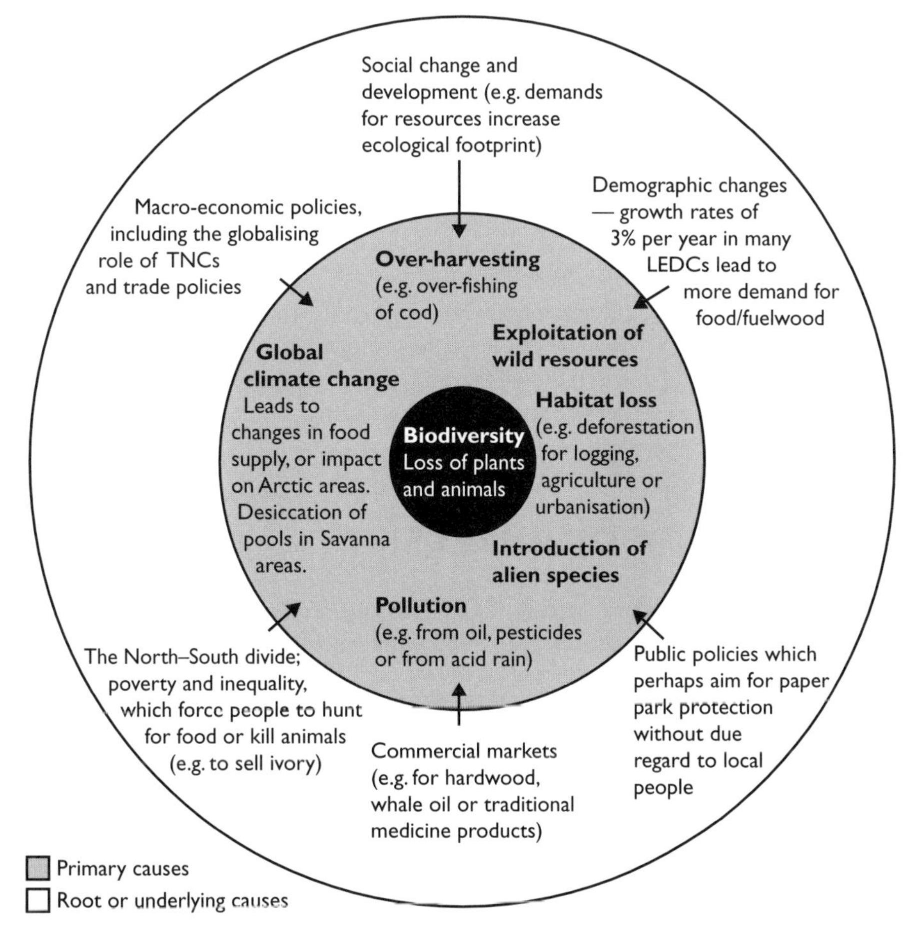

Figure 35 Threats to biodiversity

Global distribution of biodiversity

The global distribution of biodiversity varies enormously.

Tropical rainforests are the most biodiverse terrestrial biomes.

100 square kilometres of typical rainforest in primary condition might support:
- 1500 species of flowering plant
- over 30 000 insect species
- 750 species of tree
- 400 species of bird
- 150 butterfly species
- 100 species of reptile
- 60 species of amphibian

Coral reefs are the rainforests of the oceans. They provide thousands of ecological niches. Land areas experiencing Mediterranean climates are the nearest terrestrial rival to rainforests, with a huge diversity of plant species but fewer species of large mammal.

Recent studies have revealed around 20 ecological **hot spots**, largely in rainforests or Mediterranean areas. A hot spot must contain at least 0.5% or 1500 of the Earth's recorded plant species. Hot spots are areas containing rare, often unique (endemic) species, which are threatened by environmental degradation. Hot spots cover just over 1% of the Earth's surface and contain 44% of the total plant species and 35% of the total animal species.

Many of the most biodiverse areas, with high percentages of endemic species, are found within LEDCs. These countries have the least money to develop sophisticated conservation strategies to manage the underlying causes of biodiversity loss. Closure of areas to protect ecosystems can threaten the existence of people who are often living below the poverty line.

Conserving biodiversity

Conserving biodiversity requires a global framework for action combined with the effective operation of national, regional and local strategies.

Until the early 1990s, biodiversity loss was looked at as a scientific global issue with a series of frameworks for **species protection** and **habitat conservation**. Measures for species protection included the Convention on International Trade in Endangered Species (CITES), which banned trade in threatened species and their products. Measures for habitat conservation included the creation of RAMSAR, which involved the conservation of wetlands of international importance.

Additionally, a number of global frameworks were developed to conserve areas of outstanding ecological importance. These involved United Nations agencies such as

UNESCO (responsible for the biosphere programme) and UNEP (responsible for global environmental monitoring systems). Some private organisations are responsible for designation and categorisation of a range of protected area categories, including world heritage sites. The WWF is a non-government organisation, which works with governments on the management of protected areas.

Figure 36 shows the enormous range of protected areas. Note the many different and frequently conflicting aims.

Key: Prime aim; Important objective; An objective where resources permit; Not applicable

Management category	Conserve and improve hydrological systems	Prevent and control erosion and sedimentation	Conserve and improve timber and related forest resources	Conserve representative sample species (protection)	Habitat conservation	Protect wildlife resources	Conserve genetic resources	Provide opportunities for recreation	Provide opportunities for research, monitoring and education	Improve/perfect environmental quality	Achieve conservation and rural reserve development	Support lifestyles of indigenous people	Promote sustainable rural development	Control exploitation of resources
Biological reserve	Important	Important	Not applicable	Prime aim	Prime aim	Important	Prime aim	Not applicable	Prime aim	Important	Not applicable	Not applicable	Not applicable	Prime aim
National park	Important	Important	Not applicable	Important	Prime aim	Important	Resources permit	Prime aim	Prime aim	Prime aim	Not applicable	Not applicable	Prime aim	Prime aim
Forest reserve	Prime aim	Prime aim	Prime aim	Resources permit	Resources permit	Important	Prime aim	Important	Prime aim	Important	Prime aim	Important	Important	Prime aim
Wildlife/wetlands refuge (RAMSAR)	Important	Important	Not applicable	Prime aim	Prime aim	Prime aim	Important	Resources permit	Prime aim	Important	Not applicable	Not applicable	Not applicable	Important
World heritage site	Not applicable	Not applicable	Not applicable	Not applicable	Prime aim	Not applicable	Not applicable	Prime aim	Prime aim	Prime aim	Important	Resources permit	Important	Prime aim
Biosphere reserve	Important	Important	Resources permit	Prime aim	Prime aim	Prime aim	Prime aim	Not applicable	Prime aim	Important	Prime aim	Important	Prime aim	Prime aim
'Managed resource' protected area	Not applicable	Important	Resources permit	Not applicable	Important	Important	Not applicable	Prime aim	Not applicable	Prime aim	Prime aim	Prime aim	Prime aim	Prime aim

Figure 36 Conservation objectives in a range of protected areas

Conservation strategies

Monitoring of ecosystems from 1970–95 by UNEP and WWF suggested up to 30% global loss of key species and habitats, despite the many global initiatives and national innovations. It became apparent that **protection-only strategies** would *not* conserve biodiversity in the long term.

Six root causes for this were identified:

- Biodiversity loss was largely tackled on a global scale using global frameworks, when the actual occurrence of a problem was highly localised, with diverse causes.
- Solutions had not worked — often for political reasons, as the institutions involved at a national, regional and local level were uncoordinated, underfunded and did not share the same aims.
- In the poorest countries of the world, the conflict between economic development and conservation had not been fully understood. In particular, conservation of

large animals could upset the food supply of a growing local population, as hunting and trapping were banned.

- The policies were narrow. They emphasised protection and failed to see that conservation was heavily influenced by social, economic, cultural and political factors.
- Although quite numerous — some 10 000 sites in 1995 — the protected areas only covered around 6% of the Earth's surface. Their distribution was very uneven by continent, with many sites too restricted in size to survive damaging actions or events.
- Many schemes were country-based, whereas ecosystems such as the rainforest crossed many national boundaries.

The Rio Summit Biodiversity Treaty of 1992 aimed to conserve biodiversity via the sustainable use of components and the fair and equitable sharing of benefits. This was reiterated at the Johannesburg Earth Summit of 2002. The main problem is to achieve a balance between growth and consumption versus sustainable development and conservation. In areas such as the Galapagos, the growth of tourism and the rich potential of the fisheries threaten to tilt the balance away from sustainability.

The spectrum of conservation strategies is summarised below:

Increasing exploitation ↓

- Total protection pre-1990s style — nature or scientific reserve with no access for local people (e.g. Costa Rican forest or biological reserves).
- Protection strategies with economic opportunities for adjacent areas, known as buffer zones (e.g. biosphere reserves, such as Korup).
- Protection strategies, which encourage sustainable development, combined with conservation and sharing of benefits (e.g. SMMA, St Lucia); providing incentives for conservation with non-timber forest products, harvest and ecotourism (e.g. CAMPFIRE); extractive reserves.
- Token protectionism combined with growth; development largely by northern TNCs; little trickle-down of wealth gained (e.g. Indonesian rainforests).

After 1990 a new model for conservation projects developed which emphasised the following:

- Conservation strategies should be holistic, with strategies involving local economic development as a conservation tool, for example, the development of ecotourism in aboriginal reserves in Australia or Ecuador or the extractive reserves in Brazil.
- Concerns about pollution and climate change indicated that larger geographical areas (bioregions) should be conserved, with interlinking areas across national boundaries and migration corridors (e.g. Peace Parks across Africa).
- A recognition that schemes in LEDCs require financial pump-priming from MEDCs (e.g. the Debt for Nature Scheme). Many LEDCs are fearful of MEDC interest, particularly of TNCs, in their gene pools.
- A global top-down framework to be combined with bottom-up schemes involving local people in valuing ecosystems, yet at the same time gaining benefit from them. It was realised that simultaneous operations on a variety of scales from local to global are required to conserve biodiversity.

- *In situ* conservation may need to be combined with *ex situ* conservation schemes, such as zoos and gene banks.

Controversial conservation issues

Selection of areas

Should the approach be to conserve a *comprehensive representative* area of each ecosystem (WWF/World Bank Global 200 ecoregions approach) or should conservation be concentrated in *threatened hot spots* (Myers), defined by species diversity, endemism or distinctiveness?

Other approaches have concentrated on specifics such as endemic bird areas, forest frontiers and endangered animal species — known as the Flagship Scheme.

The 2002 Convention on Biodiversity (183 signatories) focused on endangered plants on the IUCN **Red List**. There are 16 clear targets for each country to meet by 2010. For the first time, these targets apply to anywhere the endangered plants grow, not only to protected areas.

Design of reserves

Reserves should be large enough to be a **minimum dynamic** area for a habitat to survive and support a **minimum viable** population. An equal area divided between several smaller reserves may be less appropriate than a single large one (SLOSS debate — single large or several small) but may be a better buffer against disease, predation or severe natural disturbance. Small reserves are more vulnerable to erosion by human activities. Sprawling linear reserves may cover more diverse habitats and allow greater species movement. The latest thinking on sustainable management emphasises the importance of buffer zones in protecting a core-conserved area. Figure 37 shows a 'cluster' biosphere reserve, reflecting the latest thinking.

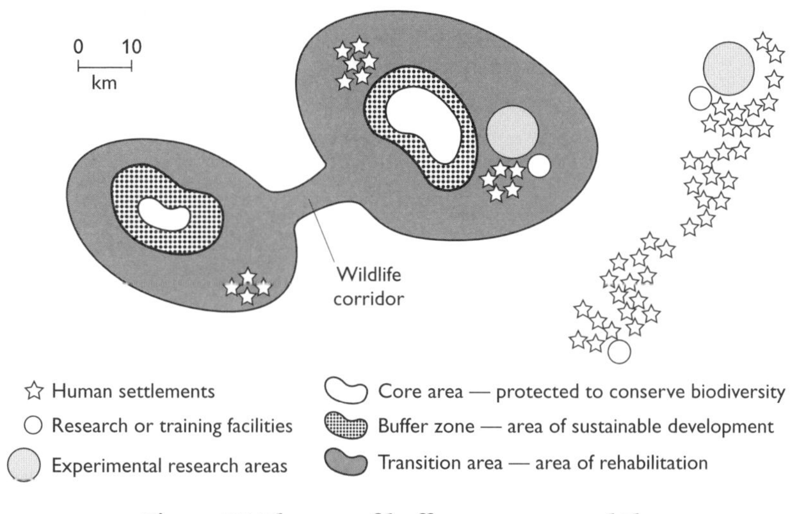

Figure 37 The use of buffer zones around the Great Smoky Mountains Park in the USA

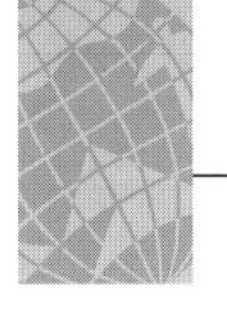

Rehabilitation and restoration

How viable is it to attempt to restore damaged ecosystems? Even small-scale schemes have high costs and many argue that replacement ecosystems have low biodiversity. However, it can be argued that new ecosystems have benefits for visual amenity and wildlife. *Ex situ* conservation (e.g. in 2003), which can produce a supply of rare species to release into the restored habitat, may be useful here.

Sustainable development of ecosystems

Table 9 shows four principles of sustainable development.

Table 9

Futurity	Environment
Present generations should leave future generations with the ability to maintain present standards of living. We are entitled to use up finite natural resources only if we provide future generations with the know-how, through science, technology and social organisation, to maintain living standards from what is left. This is the Brundtland principle.	We should seek to preserve the integrity of ecosystems, both at the local level and at the scale of the biosphere, in order not to disrupt natural processes essential for safeguarding human life and for maintaining biodiversity. This includes the ecofriendly management of water, land, wildlife and forests.
Public participation	**Equity and social justice**
The public should be aware of, and participate in, the process of change towards sustainable development in line with Rio Summit (1992) Principle 10. Environmental issues are best handled with the participation of the citizens. Each individual should have the information and opportunity to participate in the decision-making process (bottom-up).	This principle implies fair shares for all. If there is a finite amount that we may consume or use, then we must share what we already have far more than is currently the case. Therefore, equality of access to the world's global resources must be the guiding principle, thus improving the lives of the poor (Johannesburg 2002).

Issues

There are a number of constraints that make sustainable development difficult:

- The most biodiverse ecosystems with high **bioquality**, such as rainforests and reefs, are found in tropical regions where the countries are largely LEDCs.
- There are tensions between short-term gain and long-term benefits, as well as local versus distance issues.
- Many of the goods provided by these ecosystems are not found within MEDCs, which have set up transnational corporations in order to exploit them. Short-term profit from goods is at the expense of services and possible future use.

Conflicting values placed on ecosystems

Goods: direct use

Tables 10–12 summarise some of the conflicting values placed on ecosystems which make sustainable development so complex.

Table 10 Direct use

Value	Description	Who mainly benefits
Subsistence	Biodiversity supports and provides products that can be hunted or gathered from natural, semi-natural or managed systems (e.g. food, fruits, building and clothing materials, fuelwood, medicines, livestock fodder, materials such as dyes, resins, gums)	Mostly rural people, often poorer groups, traditional peoples and subsistence farmers Often they have no formal ownership of land
Tradable (commercial)	Biodiversity supports and provides a range of products, which can be traded in markets (e.g. bush meat or bif, crops, timber, fish and genetic resources)	Small-scale and large-scale commercial enterprises including MNEs and their employees (e.g. collectors, artisans, plantation workers)

Table 11 Services: indirect use

Value	Description	Who mainly benefits
Environmental services	Biodiversity is the medium through which air, water, gases and chemicals are cycled and exchanged to create environmental services (e.g. watershed protection, wildlife habitat, carbon storage, pest and disease control and soil formation)	At the local level, small-scale producers rely heavily on local environmental services (e.g. nutrient cycling) At global levels, all benefit, especially rich nations
Research services	Biodiversity equals genetic diversity, which enables scientists to create new varieties of seeds and drugs	Commercial farmers; plant and animal breeders; researchers and scientists; agrochemical, food and pharmaceutical companies

Table 12 Future use

Value	Description	Who mainly benefits
Non-uses, i.e. leaving the ecosystem as it is	Biodiversity may hold species or genes that could help insure against future risk (e.g. combating new diseases or ensuring adaptability to climate change)	Future generations
Conservation	Biodiversity holds an intrinsic worth, which transcends its uses and financial values (e.g. aesthetic enjoyment via ecotourism)	Urban dwellers, photographers and conservationists; tourists and tourism companies

In general, forces of **market liberalisation** (by free trade), **globalisation** (by multinational operations), **population growth** (increased pressure on rural land), **increasing global levels of affluence** (demands for resources) and **urbanisation** (increased need for resources to be exported out of the region) have encouraged commercial exploitation at the expense of indigenous use, which tends to be regarded as less valuable or important.

Commercial exploitation stands in the way of sustainable development and usually leads to increasing inequality between rich and poor. Trickle-down of wealth frequently does not occur on a local scale. Equal access for the poorer people (**equity**) and for future generations is being jeopardised. Figure 38 illustrates this inequality.

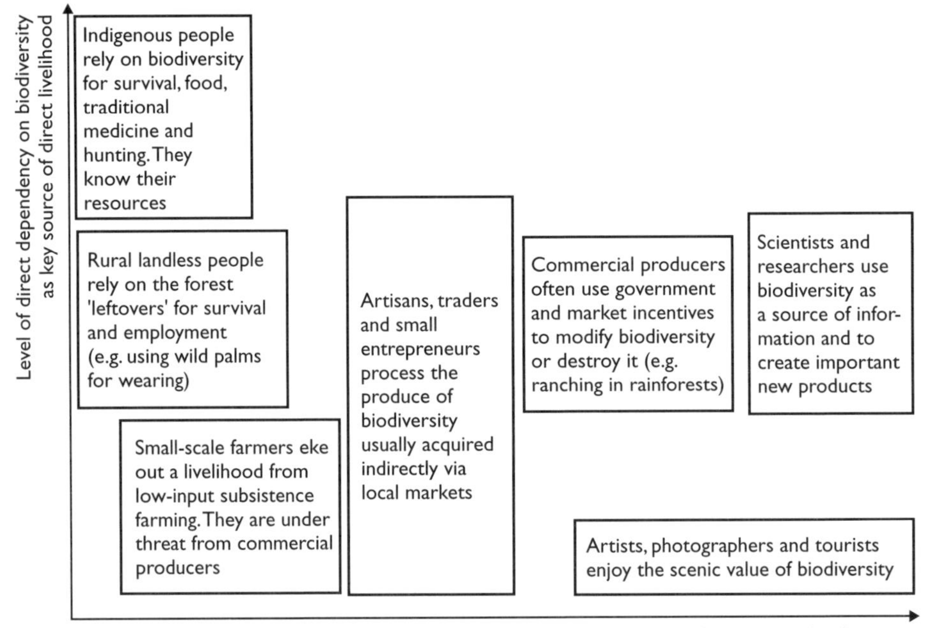

Figure 38

People directly dependent on biodiversity for all their needs, such as indigenous peoples, are most at risk of losing their livelihood. It is very difficult for these groups to be involved in sustainable development when they do not even possess maps to show their ownership rights.

Sustainable management requires a tremendous amount of preparation, for example in establishing ownership of the resource, mediating between various stakeholders whose uses may conflict and establishing a framework to empower local people to develop the resources. **Africapacity** is a major new multinational project, which is building up the organisational capacities of local NGOs.

While improved production technologies (e.g. lean and clean) may help to address some issues of resource use, biodiversity is still modified and pressure continues to be put on the environment.

Questions & Answers

In this section of the guide there are three questions based on the topic areas outlined in the Content Guidance section. Each question is worth 25 marks. There is also a cross-unit question, worth 30 marks. You should divide your time according to the mark allocation.

To answer at A2 you must be prepared to assemble your response into coherent prose consisting of sentences and paragraphs, with a brief introduction and conclusion. Most questions will be in two parts. You are expected to use the information, map, table or cartoon supplied with the question as a stimulus for your answer.

You might be asked to describe, but at this level the question is more likely to ask you to 'explain', 'state how and why', 'give reasons for' or 'evaluate'.

Examiner's comments

Candidate responses are followed by examiner's comments. These are preceded by the icon *e*. The comments indicate how each answer would have been graded in the actual exam and give suggestions for improvement.

Questions at A2 involve assessment using a system of 'Levels'.

Level 1

The answer lacks breadth and is vague. There is little supporting detail. Geographical terminology is lacking or basic. The candidate might not be answering the question set, but rather one that he or she had hoped for.

Level 2

There is some breadth and/or depth to the answer, but it might be unbalanced and not follow up certain aspects. The command words in the question are noted by implication and in the conclusion. The answer describes rather than explains when requested.

Level 3

The answer has both breadth of coverage and depth of understanding. It is relevant, precise and answers in a logical fashion. The candidate does exactly what the command words require.

Quality of written communication

Up to 4 marks are added to the paper *total* for the following qualities:

- the structure and ordering of the response into a logical answer to the question
- appropriate use of geographical terminology
- punctuation and grammar
- the quality of spelling (if you are dyslexic, seek special consideration)

Question 1

Changing weather

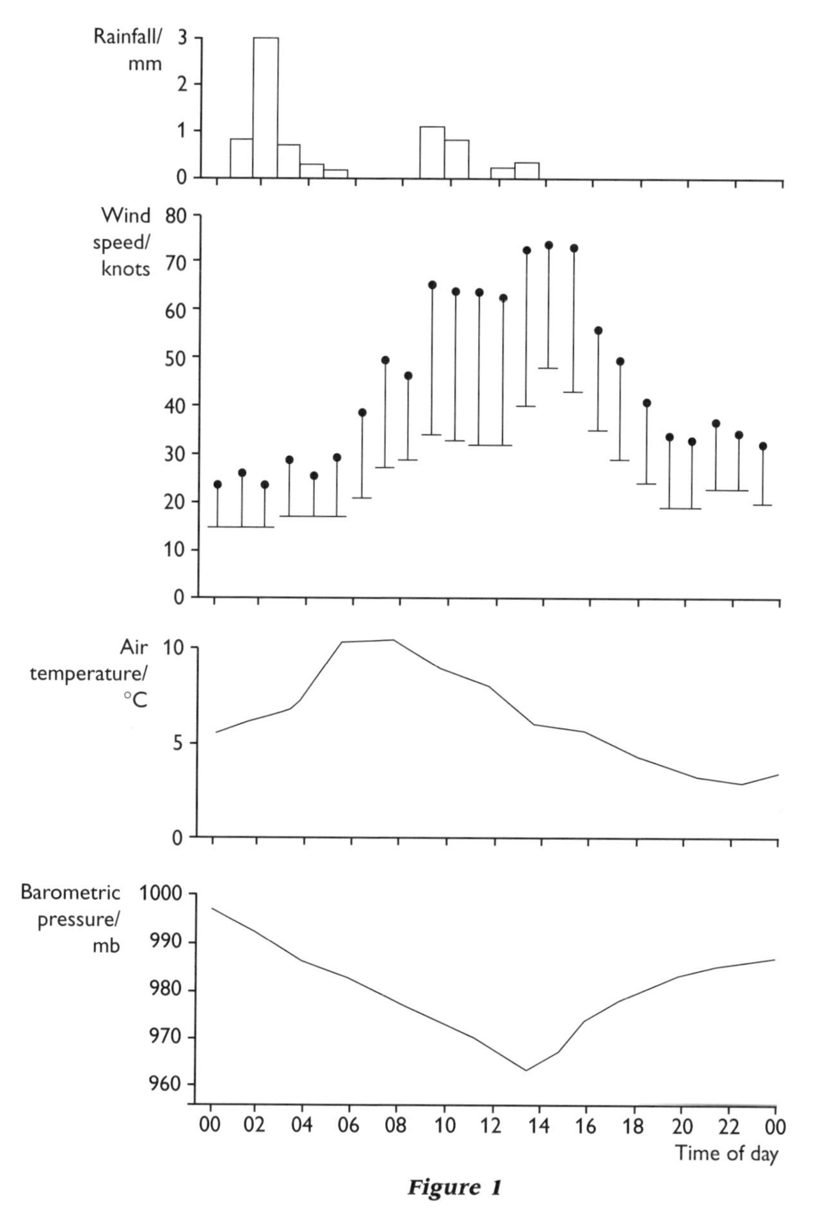

Figure 1

Study Figure 1, which shows weather conditions in Oxford during the passage of a depression on 25 January 1990.

(a) Suggest reasons for the changes in the weather conditions shown. (12 marks)

(b) Explain why depressions can sometimes cause severe weather. (13 marks)

Total: 25 marks

Answer to part (a): C-grade response

(a) A depression is a low-pressure system formed by the meeting of two contrasting airmasses at the polar front. Therefore, it has fronts. The onset of the depression can be seen from the falling air pressure. As the depression passes over Oxford, the air pressure begins to rise again. The following sequence is always found in a non-occluded depression: warm front–warm sector–cold front, and this can be clearly seen in the air temperatures. Precipitation is always linked to the fronts as uplift of warm, lighter air occurs, forming frontal rainfall. Two episodes are shown, the first at the warm front being the larger amount. As depressions are low-pressure systems, air is drawn towards the lows and the change in pressure can lead to very gusty winds. Gales are likely as the depression passes, usually from the northwest.

e This is a Level 2 response, which scores 7 or 8 marks. The coverage is sound with some explanation. To achieve Level 3, there should be more explanation, for instance looking at contrasting frontal patterns. A simple cross-sectional diagram of a depression would help here. There is also no precise data support, for example deepness of the depression could explain the gustiness and strength of the winds. For a Level 3, grade A answer, the examiner is looking for the following information:

- There are two periods of rainfall: 6 mm after 1–8 hours and around 3 mm over an extended period (9–14 hours), likely to be associated with the passage of the warm front and the cold front. Details of frontal uplift should be included, illustrated with diagrams.
- Windspeeds reach a peak of up to 70 knots. High wind speeds are associated with severe pressure gradients caused by tightly packed isobars, as winds are drawn towards a deep low pressure in a cyclone vortex. A diagram would be useful here.
- Air temperatures initially rise from 5 to 10°C, reflecting the arrival of the warm front. They are maintained at 10°C (warm sector) before falling dramatically to around 3°C as the cold front occurs.
- The air pressure shows a steady fall to 966 mb after nearly 14 hours, and a steady rise to 978 mb as the depression passes.

To ensure good structure in a data-stimulus answer, annotate the resource and work out some precise measurements (e.g. rainfall amount) before you begin. Remember this question does not ask specifically for a description. However, you are expected to quote data to support your answer.

■ ■ ■

Answer to part (b): C-grade response

(b) The main type of severe weather brought by depressions is very strong winds, especially after the cold front when northwesterly gales occur. The deeper the depression is, the stronger the winds are. These winds can whip up very strong waves and when combined with a storm surge and an onshore direction can lead to coastal floods, like in Towyn. Depressions can also bring heavy rainfall — often associated with warm fronts. If antecedent water levels are high, this can lead to strong floods (e.g. Banbury, 1998). Just occasionally, when you get violent uplift at the cold front so that the cold air wedges underneath some really warm air, depressions can lead to heavy thunderstorms and hail. In general, the greater the contrast in temperature of the airmasses, the more damage a depression can do.

e This is a Level 2 response, for 7 or 8 marks. Some good understanding is shown of variability in depressions but the examples are used rather casually. It is rather brief, with no mention of snow and insufficient detail on winds. Like many C-grade answers, it is on the right lines. More development of linked explanation and supporting examples are needed for an A-grade. Answers should include details of:

- strong winds and gales, associated with a deep depression
- associated coastal storm surges and storm waves
- very heavy rainfall and floods associated with slow-moving depressions
- the possibility of hailstorms because of violent uplift at the cold front
- occasional lows that draw in a very cold, Arctic area leading to severe snow falls which, combined with strong winds, give rise to blizzards

Global warming

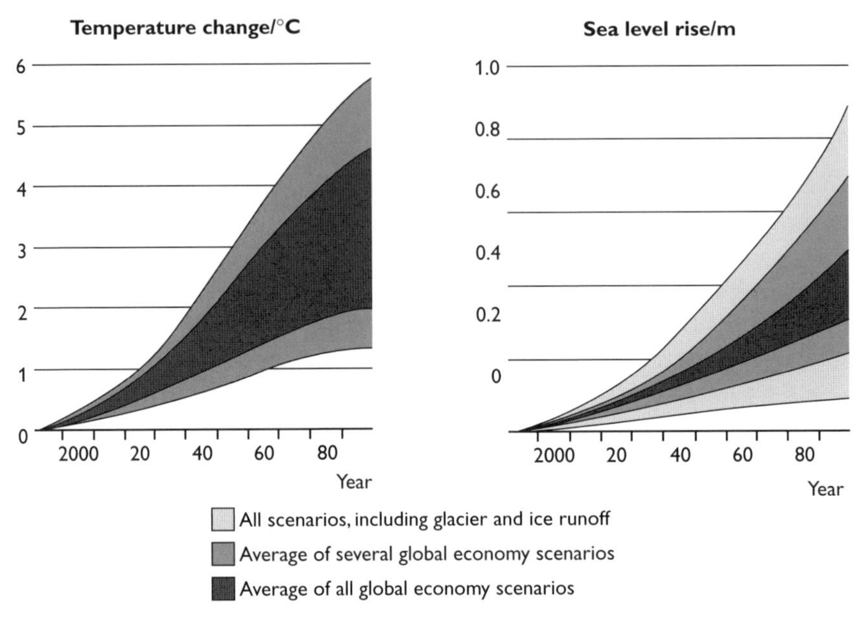

Figure 1

Study Figure 1. It was developed by IPCC and shows possible scenarios for temperature change and sea level rise as a result of global warming.

(a) Explain why scientists find it difficult to predict temperature changes and sea level rises. (12 marks)

(b) Assess the likely impact of temperature change and sea level rises on the environment. (13 marks)

Total: 25 marks

■ ■ ■

Answer to part (a): C-grade response

(a) There are two main reasons why it is difficult to predict the rate and impact of enhanced global warming, which is thought to be responsible for temperature and sea level rise.

First, the amount of global warming depends on the amount of greenhouse gases emitted. These are a mixture of carbon monoxide, carbon dioxide, oxides of nitrogen, ozone, methane and CFCs. The amounts are linked to levels of development in many countries — how much they will industrialise; how much energy they will use; how intensive their farming will be. The greenhouse gas output controls the rate.

The second key issue is what the world and the countries are going to do about it, and how successful they will be in controlling greenhouse gases. Summits such as Kyoto and The Hague have developed protocols with individual emission targets for carbon dioxide for all the world's countries. However, many key players such as the USA, Japan and Canada have not signed up because of worries that it will inhibit their economic development. A complicated system of carbon credits and carbon sinks may not be successful. If the world does reach global agreements, they will be long term and slow to operate, so it is very difficult to predict temperature rises.

Sea level is a knock-on issue. It is temperature rise that has an impact on how fast oceans warm and expand and also how quickly the ice caps melt, contributing to a worldwide rise in the level of the oceans. In some places, isostatic recovery is taking place, so the relative rise will be less.

This is a good Level 2 response, for 9 marks, showing sound understanding of the main issues. However, to gain a grade A the answer should:

- give more detail of the processes and the gases involved. Carbon dioxide is the only one mentioned in any detail.
- use more geographical terminology
- link to the resource to discuss reasons for the contrasting scenarios

■ ■ ■

Answer to part (a): A-grade response

(a) Global warming results from an enhanced greenhouse effect brought about by increased greenhouse gases. It is difficult to predict human impacts. Increased intensification of land use could lead to increases in methane. Increased industrialisation could lead to more nitrogen oxides and carbon dioxide. Increased urbanisation and affluence could lead to more energy use and car ownership — hence rising carbon dioxide levels. It is even more difficult to assess the impact of CFCs because this involves past actions and their effects can last for 180 years. Rates of development are hard to predict, as are rates of slowdown as a result of global legislation and action following Kyoto. The impacts of local and national sustainable energy and transport plans are also hard to predict. Complex systems of carbon credits and carbon sinks are difficult to monitor. The increase in greenhouse gases is possibly because of fast-growing large economies such as China and the reluctance of MEDCs such as the USA and Japan to kerb carbon dioxide emissions. Some countries are making excellent progress in controlling greenhouse gases; some have LEDC economies that barely have an impact.

Sea level rise is linked to global warming but is related to thermal expansion of the oceans (directly linked with temperature) and to the unpredictability of the rates at which ice caps in Greenland and Antarctica are melting, leading to eustatic rises in sea level.

Answer to part (b): C-grade response

(b) I have chosen to look at the environment with respect to the British Isles. Here there is a northwest–southwest divide in terms of climate change. The northwest of the British Isles will get both warmer (possibly 2°C) and about 20% wetter, with much stormier winters. The southeast will get hotter (2–3°C) and drier in the summer, just like the Mediterranean.

This will cause a major hydrological problem with potential droughts and huge bills for subsidence because of shrinking. Since the length of the growing season will increase, this will make UK trees grow quickly, with higher yields for forestry. Some arctic species such as tundra plants and rare birds such as ptarmigan will disappear from northwest Scotland. However, new species will arrive in the south, particularly more butterflies and other insects.

Rises in sea level and a higher incidence of storms will result in coastal flooding of large areas of fenland and increased coastal erosion of soft rock coasts like Holderness. This will require billions of pounds of expenditure on sea defences.

e This is a high-risk strategy. The national answer was quite well learnt with accurate information on the likely changes and impacts of both temperature increase and sea level rise, but the decision to limit the scope of the answer made it rather brief. It lacks the range needed for an A-grade. It also fails to develop the bigger picture of climate belts shifting, although the migration northwards was hinted at by the mention of the Mediterranean. For this reason, it scores 6 marks. Direct impacts include:

- migration of climatic belts (estimated 150–300 km) and the subsequent impact on the hydrological cycle and ecosystems (both plants and animals). The impact of humidity changes is also significant. You should include locational detail and try to strike a balance between winners and losers.
- increased unpredictability of weather. Warming is possibly linked to El Niño or other oscillations occurring more frequently. You should include the impact of greater extremes, such as the possibility of more hurricanes in the Caribbean.
- increasing ocean temperatures can lead to thermal expansion, which indirectly leads to rising sea levels. This impacts on ecosystems (e.g. widespread coral bleaching).
- increased spread of some pests and diseases (e.g. malaria).

Indirectly, there is increased coastal erosion leading to the need for defences. Rising sea levels impact on coastal lowlands and deltas, causing flooding. There are problems for small island atoll economies such as Tuvalu.

e **Overall, the C-grade response was a little unbalanced, with part (a) significantly longer and better than part (b) in spite of similar mark weighting.**

Ecosystem conservation and management

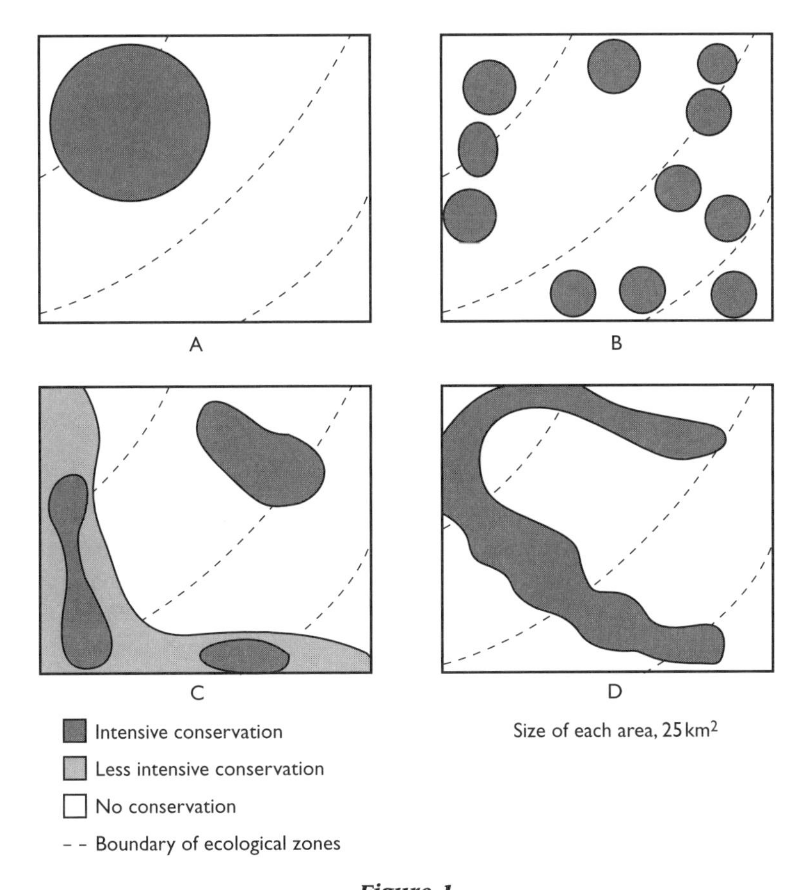

Figure 1

(a) Study Figure 1, which shows four designs for an ecological reserve. Assess the relative strengths and weaknesses of options A, B, C and D. (12 marks)

(b) With reference to one named global ecosystem, assess the extent to which conflicts occur between exploitation and conservation. (13 marks)

Total: 25 marks

■ ■ ■

Answer to part (a): B-grade response

(a) The four options all have some strengths. A is a large reserve which gives good protection but only covers two zones. There might be high-value species outside. B has good coverage of the ecological zones, so the maximum biodiversity will be conserved, protecting against diseases and pests, as in some zones there is a back-up reserve. However, these small fragmented zones may be easily encroached on by farming and they will be difficult and costly to monitor and police against illegal hunting. C consists of two reserves and covers all the ecological zones. The idea of a less intensive conservation zone around the larger reserve is a good idea, because it will protect the main reserve and presumably allow activities such as ecotourism that will give the local people extra money. D is the largest area with a very long boundary. It will be the most costly to manage. It does cover all the zones. Its linear shape could allow migration, if there was a problem such as global warming.

e This is a Level 2 answer, scoring 8 or 9 marks, which shows sound data analysis and uses the resource effectively to consider issues of size and coverage. However, although there is some understanding of design, it lacks the terminology and theoretical perspective of ideas, such as buffer zones and the SLOSS debate, required for an A-grade.

■ ■ ■

Answer to part (a): A-grade response

(a) Option A covers only two zones — hopefully of the highest biodiversity. It is around 4 km² and this larger size may enable it to survive natural disturbances and be a better buffer against disease. Even so, it is still small and two ecological zones are not covered.

Option B consists of 11 mini-areas and covers all ecological zones (representative species). However, at around 0.5 km², fragmentation will make these reserves unviable and subject to human erosion by hunting and agricultural development.

Option C covers all the zones with a large area surrounded by some sort of buffer zone, where presumably sustainable development can take place. This could mean far more effective protection of core areas.

Option D is linear and covers all zones. Its shape will encourage effective migration of species in the event of climate change. It is a single, large reserve, only 0.5 km wide at its narrowest, so could easily be destroyed (e.g. by hunting).

■ ■ ■

Answer to part (b): B-grade response

(b) Coral reefs are one of the most productive ecosystems in the world. They are also some of the most exploited ecosystems, because of their high biodiversity. Reefs provide shelter for an enormous range of organisms. As reefs occur in tropical countries, most of which are poor LEDCs and are often small islands, reefs provide a vital form of subsistence for many people. Usually their actions damage the reef, rather than totally destroy it. However, using the reef for limestone and cement (the Maldives), selling the coral (e.g. black coral in Jamaica) or dynamiting it to catch fish (Philippines), is destructive. As tourists queue to visit reefs (e.g. in Hawaii), reefs can be heavily damaged by boat anchors and divers' feet. This means there is a conflict because the reef then looks less attractive.

Exploitation of fish can be damaging, as it can upset the food web, leading to a loss of biodiversity. The Far Eastern market for fish has led to over-exploitation of many Pacific reefs.

Over-exploitation of land close to coral reefs causes major problems of siltation and pollution, both of which can destroy corals. In particular, in St Lucia deforestation of the watersheds for commercial agriculture leads to huge volumes of silt being washed down on to the reefs (e.g. by tropical storm Debbie), which blocks out the light needed for algae to photosynthesise.

As the reefs have such high biodiversity, conservation is vital. One way to do this is to create marine protection areas, so that there will be no exploitation. This requires large sums of money to police, as local people rely on the reef for subsistence, particularly for fish. In St Lucia, the creation of the marine reserves caused conflict with the fishermen at Souffrière. The only solution was the creation of sustainable marine management areas that included fishing priority areas and marine ecotourism (divers' fees paid for the policing). As local people were involved in the management, it worked. The fish multiplied in the protected environment.

e This is a very good Level 2 answer, for 9 or 10 marks, with well located examples, sound knowledge of exploitation and conservation, and some mention of conflicts. A Level 3 answer needs more 'spelling out of the conflicts' and more assessment of their extent.

e **As is the case with many low B-grade responses, the material in this answer was entirely appropriate (genuinely global). However, to achieve the top grade the material needed to have been used in a more focused way, especially in part (b). The question asks for a named global ecosystem, so rainforest would also be a good example. You would gain a maximum of 9 marks for using an example from one continent, such as Korup, and a maximum of 5 marks for a small-scale area such as a local woodland.**

■ ■ ■

Answer to part (b): A-grade response

(b) Exploitation of commercial timber, usually by TNCs, causes conflict because it destroys goods such as fruits, nuts and plants with medicinal properties that are used by indigenous peoples, who often have poorly documented ownership of the land. It also has a major impact on the forest environment (hydrology, soil erosion), which means that there is a loss of both the vital services that rainforests provide (carbon sinks, green lungs) and the future use of the gene pool. However, sustainable forestry (e.g. in Malaysia) can be practised, although reafforestation takes a long time to be effective because of the time needed for hardwoods to mature.

Exploitation for alternative land uses such as cattle ranching in Central America and Brazil, mineral exploitation, HEP development, and the building of infrastructure (e.g. the transamazonian highway) which led to subsequent increased settlement, can also hasten deforestation and degradation. Alternative uses such as rainforest tourism, which is usually low-volume ecotourism, cause limited conflicts.

Exploitation by native peoples for subsistence can lead to forest degradation, as the secondary forest replacing fuelwood and slash and burn is of lower quality.

Exploitation of specific plants and animals (e.g. parrots as pets) can have a disastrous effect on conservation and may need ex situ support.

If a total protection strategy is adopted, conservation can cause conflicts because of exploitation by both indigenous peoples and commercial users. This leads to illegal hunting and felling. More usually, some provision is made for economic development by indigenous peoples in forest zones (e.g. the extractive reserve concept in Brazil), creating sustainable forest usage (e.g. ecotourism in Equador), and developing buffer zones around forest reserves, in which local people can hunt and gather (e.g. Korup).

Cross-unit question

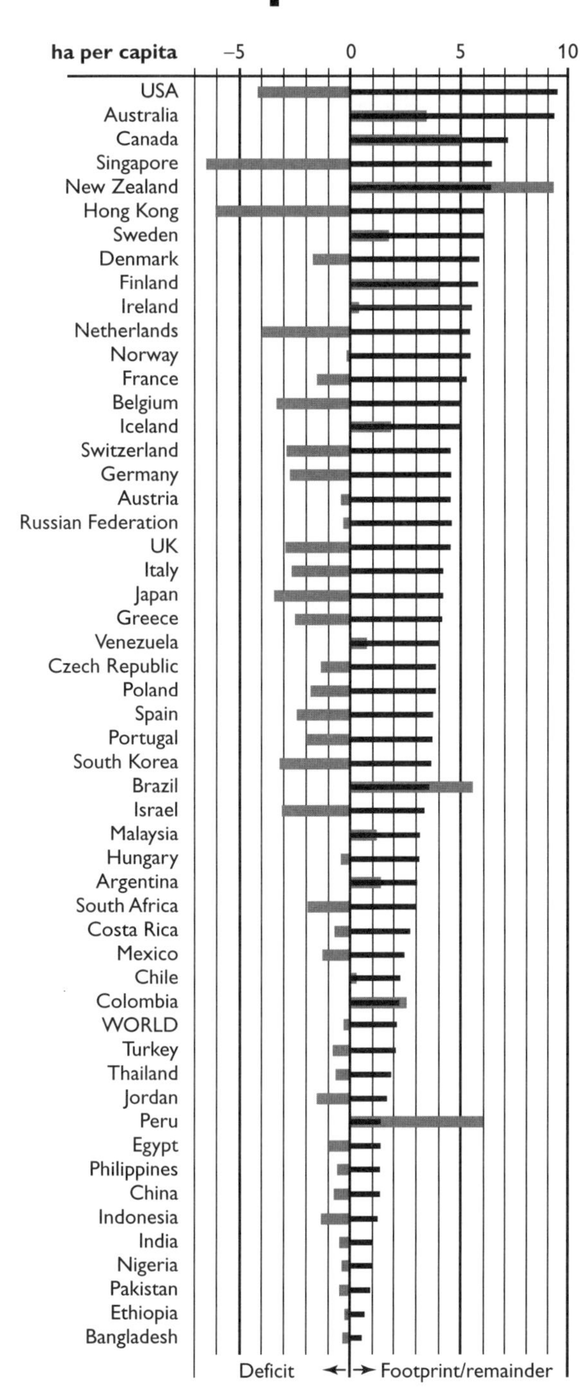

Figure 1

The black line indicates the average per capita ecological footprint. The grey line indicates the extent to which the country could meet its needs from within its own bioproductive capacity, which includes land and sea. If the grey line extends to the right, it indicates a national ecological remainder. If the grey line extends to the left, it indicates a national ecological deficit.

question

Study Figure 1, which ranks the ecological footprints of selected countries.

(a) Describe and suggest reasons for the relationships shown between ecological footprints and bioproductive capacity. (15 marks)

(b) Explain how a country could take steps to reduce its ecological footprint. (15 marks)

Total: 30 marks

■ ■ ■

Answer to part (a): B-grade response

(a) The diagram shows a gentle decline in ecological footprints from the USA at the top to Bangladesh at the bottom. There is a straightforward correlation between ecological footprints and wealth. The more developed MEDCs can afford to use more resources, either from within the country or through trade. Apart from Hong Kong and Singapore, the top 20 for ecological footprints are all MEDCs. The bottom ten countries are all LEDCs, although China and India, with their rapid industrialisation and emergence of NICs, have a surprisingly low footprint. Interestingly, the list contains no oil-producing countries such as Saudi Arabia. The world seems to have a footprint that is slightly below the average of the countries shown.

Bioproductive capacity represents a country's ability to provide resources, such as forests, and land for farming and infrastructure. There is no link with development, so all sorts of relationships are shown. Around 70% of countries have a deficit — as does the world overall — which is a matter of real concern. Around 30% of countries shown have a remainder — presumably they have lots of resources such as forests, building land, minerals and possibly clean power such as HEP — so they rely on producing lower fossil fuel emissions. An example is New Zealand.

e This candidate shows good analysis of the ecological footprint variations, with sound understanding, although fails to give a definition. There is good use of both examples, showing understanding and giving reasons for the overall relationships. Bioproductive capacity is understood but the deficit is not fully explained. There is no mention of issues of population to resource ratios, country size or population density. The remainder of the answer shows some understanding, with limited examples. This answer scores 10 marks.

■ ■ ■

Answer to part (a): A-grade response

(a) The ecological footprint is a measure of the pressure placed by humans on ecosystems. The four major components are land for building, land to produce food, land to provide resources such as wood, and space to absorb carbon dioxide emitted from fossil fuel use and waste.

In general, the size of the footprint is directly related to the level of development. To a lesser extent, it can also be related to population size. The world's ecological footprint represents an average of the constituent countries, but in this diagram comparatively few LDCs are listed.

The relationship between ecological footprint and productive capacity — the ability of the country to produce land for food, forests, to provide resources and space for building and waste disposal — is much more complex.

There are many more countries in deficit than there are with a national remainder. The largest deficits are MEDC countries with high-density population for the land area (e.g. Singapore, Hong Kong, Netherlands, UK, Japan). These countries have little space for building or food production. Some larger countries with less consumption have smaller deficits. Deficits occur at all levels of development.

There are 13 countries with a national ecological remainder. They tend to be large countries that may have great forest wealth, such as Brazil, Finland, New Zealand, Australia and Canada. In some cases, the bioproductive capacity is well in excess of the ecological footprint (e.g. Peru).

Overall, the world is in deficit, which is an issue for concern.

■ ■ ■

Answer to part (b): B-grade response

(b) Achieving a reduced ecological footprint is all about sustainable development, so that fewer resources are used and they are used in a more environmentally friendly way. If we look at the forest footprints, one way of reducing these is to carry out more sustainable forestry projects. In Malaysia, there are careful plans to fell trees and to replace them with new forests. Other measures include the avoidance of clear felling and erosion control strategies. The Forestry Stewardship Council has supported these schemes by marketing sustainable timber to MEDC producers. Another scheme involves MEDCs paying off LEDCs' debts in return for these countries saving certain acreages of forests.

In terms of the more sustainable use of other resources, sustainable fisheries schemes have become common. In St Lucia, there is a sustainable reef management scheme that has involved the local community as stakeholders in the management of marine conservation areas, and fishing priority areas. The creation of the reserve, which stopped reef fishing, has led to a better size of fishing catch, via fish aggregation devices.

To reduce the carbon dioxide footprint, there are a number of strategies that can be employed. The largest contribution to the carbon dioxide footprint is the internal combustion engine. Therefore, countries need to establish green transport policies that rely on public transport and to develop the use of alternative cleaner, more sustainable energy strategies such as wind power. These strategies are used in both Denmark and Holland. The use of scrubbers on chimneys cuts down carbon dioxide emissions from factories.

In order to reduce the waste footprint, there has to be a strong emphasis on the development of recycling schemes for the disposal of all types of waste. Also, strategies are needed for the treatment of hazardous waste, such as CFCs from fridges.

The overuse of land for building can be controlled by sustainable urban strategies, which persuade people to live in new-style compact cities as opposed to unsustainable cities such as Phoenix, Arizona.

One aspect that is neglected is that many MEDCs trade in order to feed themselves and therefore import food from all round the world. The concept of food miles can be used to highlight unsustainability. A good strategy is to reconnect people to local farm products, thus cutting down food miles. This is part of the latest EU Common Agricultural Policy planning. Therefore, there are a number of ways that, together, could reduce the ecological footprint.

e The candidate has a fair range of strategies supported by some relevant examples. It is a well-structured response around footprint components. To gain a grade A, the answer would need to concentrate more on national strategies and the role governments can play — internationally to develop a legislation framework and nationally and locally to develop and support sustainable policies. However, this is quite a good cross-unit standard of performance, for 11 or 12 marks.

There is a range of possible suggestions, focused around policies for more sustainable use of land and resources, such as sustainable forestry, fishing, farming methods and urbanisation strategies.

Another important strategy is the use of more sustainable resources for energy and transport, in order to reduce both the reliance on fossil fuels and the emissions from them. You should include details of alternative energy strategies and green transport plans.

You could include details of international, national and local development of more sustainable policies and pro-environment laws, for example in recycling waste. You could also consider sustainable population strategies to control birth rates, to decrease the population to resource ratio.